# GEOMETRY
## IN NATURE

# GEOMETRY IN NATURE

Exploring the Morphology of the Natural World
through Projective Geometry

John Blackwood

Floris Books

*To Norma, for her unstinting support*

Published in 2012 by Floris Books

© 2012 John Blackwood
John Blackwood has asserted his rights under the
Copyright, Designs and Patents Act 1988
to be identified as the Author of this Work

British Library CIP Data available

ISBN 978-086315-921-3

Printed in Poland

# Contents

*Consider the lilies, how they grow; they neither toil nor spin; yet I tell you, even Solomon in all his glory was not arrayed like one of these.*
(Luke 12:27)

# 1. Introduction

*Figure 1.1 Machinery [Samantha Collins]*

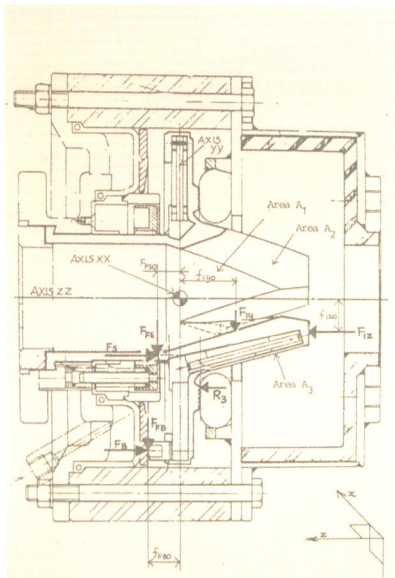

*Figure 1.2 Flow control device*

The premise for this book is that if we can have thoughts about things, then the thoughts must be inherently in the things. This is not a new idea. It may not be so easy to discover exactly what these thoughts are, but that need not be the fault of the things or the thoughts – it may be our own shortcoming.

We might not necessarily find the relevant thought that relates to the phenomena in a day, a year or even a century, but that does not mean it is not there to be found. Something guides, builds, designs and sustains the things we see, whether we choose to acknowledge it or not. Eugene Wigner, a physicist, wondered – as many have – at the curious relevance of mathematical thought to the world. He wrote an article 'The Unreasonable Effectiveness of Mathematics in the Natural Sciences' in 1960. For me, it would be unreasonable if it *weren't* effective. Rudolf Steiner (1861–1925) in *The Philosophy of Freedom* thought that we would only *begin* to find reality when we did find such thoughts. Clerk Maxwell (1831–79) found the laws of light propagation through his mathematical ability. The empirical validation came *afterwards*. However, Immanuel Kant (1724–1804) thought such reality could never be found.

## 1.1 The thought in the machine

We have little difficulty in realising that the thought is within the thing in machinery of our own making, because we put it there in the first place and otherwise it wouldn't work. Does the rest of the world somehow work that way too? And if not, why not? The fact that we can rearrange some features of the world through our own mental conjurations demonstrates that there is something within us that can find a place outside us.

We do not, I believe, need to be limited to William Shakespeare's 'pale cast of thought'. Thoughts may begin pale and be such for a long time but that need not be where they remain for us. Sometimes an idea can get us quite excited – people will even die for an idea or ideal. We only have to look at the virtual apoplexy that erupts when the word 'intelligent' is applied to the world, let alone when an ineffable *wisdom* is purported to be in nature. So much for 'pale cast'!

The 'world is an intelligent project' said Pope Benedict XVI, and in this I happen to agree with him. But that does not mean I remotely subscribe to either the fundamentalist creationism view, nor to materialistic ideologies. It is hugely unfortunate that the word 'science' has been (like so many words) captured and restricted. 'Science' means knowing, not just material knowledge. Every day we regularly use our 'non-material' knowledge of mathematics, counting and geometry. The word 'science' has been hijacked by material physical science. Such an artificial boundary is at best only a limitation; at worst it is an ideology with no more absolute credibility than any other belief structure. Jos Verhulst in his work on biology is at pains to point out that much that purports to be science is little more than an insidious ideology. Speaking of 'problems inherent in the Darwinian view [that] are well known by specialists', he says, 'In my view, this systematic disregard for legitimate objections amounts to mass indoctrination.' (Verhulst *Developmental Dynamics*, p.360).

This work is not a discussion of epistemological niceties, but tries to explore some areas where there appears to be a nexus between what we think and what we can perceive. This is, of course, the intent of genuine science. My belief is that science is the weaving between the world of ideas and the world of

*Figure 1.3 A cognitive weaving between two apparently separate worlds [Sarah Edmondsen]*

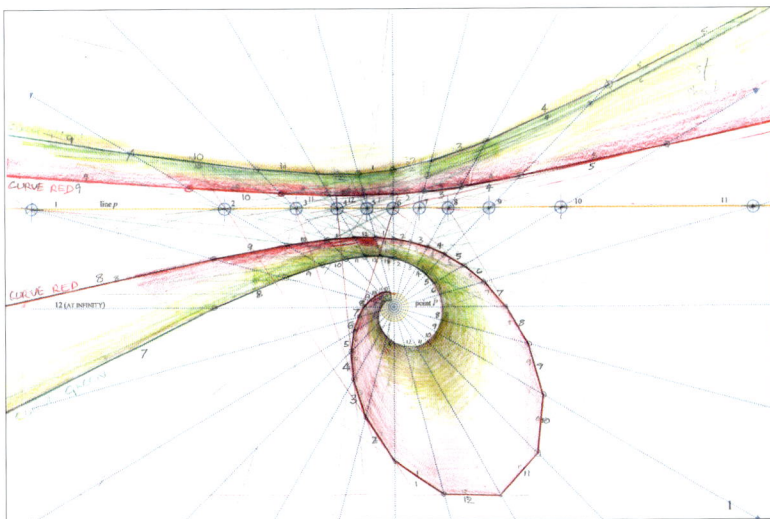

*Figure 1.4 Point-line pair in action in an ordered field*

phenomena. This is the cognitive model I adopt here. Both these directions must be given suitable weight, despite their qualitative differences, for the simple reason that they are two sides of the same thing.

Our starting point in this book is geometry – the realm of pure ideas. We start with *projective geometry* as, unlike Euclidean geometry, this geometry is not dependent on measuring. Measure and form emerge from simple transformations of projective geometry. In this book we will go on to see whether the forms we find in nature are accurate reflections of these geometrical forms.

The fundamental elements of geometry are *points*, *lines* and *planes*. Our everyday view of the world tends to see the point as the most important, with lines and planes made up of a series of points. However, I have often wondered whether we might consider the constituents of our world not just as aggregates of units, in other words point-wise, but as point-line or line-plane *pairs,* or even point-line-plane *triples.* It is not too difficult to see how a *surface* can be described through the application and motion of such a compound element. For instance a 'field' is shown in Figure 1.4 that is based on the ordered and inherent activity of a point-line pair and not just points. (The construction of this is covered in more detail in Chapter 8 on path curves.)

## 1.2 Forms of nature

The aim of the rest of this chapter is to characterise some of the forms of nature. Despite the fact that we are familiar with many of the forms we see about us, it is just this familiarity that allows us to miss something essential. Mere recognition is not a real understanding and comprehension. Are there ways to comprehend

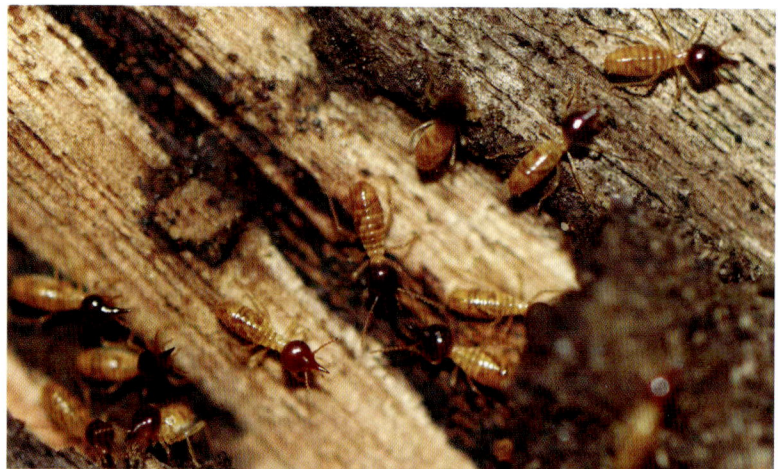

*Figure 1.5 Two bunya trees in New South Wales*
*Figure 1.6 The branching forms of the palm*
*Figure 1.7 Brush tailed possum showing bilateral (left-right) symmetry*

*Figure 1.8 Exposed termites (white ants)*

the architecture of nature? Can we begin to discern something of the systematic in the forms we see?

Here are some examples to remind us of this vast variety of forms. What, for instance, guides the form of the periphery of the great bunya trees? What orders the sweeping branches of the palm?

Equally, what organises the bilateral symmetry of the mammal's head, and how is it that this form of symmetry is so ubiquitous in animal, plant and mineral worlds? We take it for granted in our human world too. Is there a fundamental layout to the termite's body plan, abdomen, thorax and head, as with the bulk of insects? Does this speak to a primal structure underlying the living kingdoms? Is there a plan in the form of the starfish, apart from the obvious fivefold figure? What is its geometry, or the geometry of sea urchins? Are these really egg forms with vertical spine points, from mouth to backside with attendant nodes, rather than spirals?

What about leaves? What kind of the veins do leaves have? Are they reminiscent of chaos theory ideas with bifurcations of all kinds? Is this how nature tracks its way from leaf point (or nodes) to leaf periphery? Is there organisation in the nodes on the stem

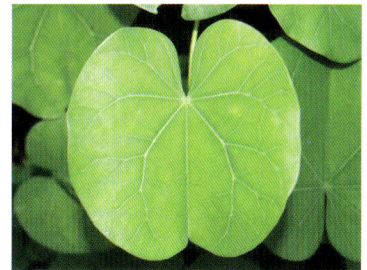

*Figure 1.9 Fivefold starfish*
*Figure 1.10 Various urchins*
*Figure 1.11 Bifurcation in leaf venation*

*Figure 1.12 Bamboo nodes*
*Figure 1.13 Alternations around main stem of a protea*

of the bamboo, and if so to what does it tend? And what of the alternation of the leaves of a protea? How often is this the case; is it typical, can it be accounted for? What about the form of the egg? Does the mysterious double spiral of the pine cone and cycad cone call for exploration? Is there something fundamentally similar in the two yet – patently – totally different?

And then there must be some systematic nature to the beauteous crystal forms. Science knows an awful lot about these from the micro side, but is there still a perspective either ignored or untouched? How, for instance does the garnet crystal come to such a full and clear rhombic dodecahedron form – even if a little bruised – rather than a messy aggregate of teensy rhombic dodecahedrons all in a molecular heap? Is it because the planar aspect has more to do with the formation of facets than we realise?

And what of the glorious animal fraternity in all their magnificence, their organ forms, the circulation structures, cellular varieties, external patterning, organ arrangement and interaction? Is there some overarching guiding archetype that incorporates all the forms and the species and their specialisations? Here are just two representatives: a magnificent red and green parrot, and the wonderful tiger in the Western Plains Zoo. Is there some form to which all the animal world approaches, yet also diverges from?

Dare we suggest that this form to which they tend, or rather *have diverged from,* is the human form? Is this the form that was the intended outcome and which all else failed to reach? Anathema though the idea may be, it has not really been given even half serious consideration by the scientific community.

*Figure 1.14 Duck's egg (with faintly visible spirals); Figure 1.15 Pine cone; Figure 1.16 Cycad cone; Figure 1.17 Quartz crystal; Figure 1.18 Garnet crystal*

Human *individuals* have names, and each human being is his or her very own unique species. In complete contrast to this, each animal species is known only by its collective name. I do not believe we are animals and I do not believe we ever were: we were always at a stage of the developing human being, not merely a surviving animal.

## 1.3 The orientation of the kingdoms of nature

Each kingdom of nature seems to have a strong relation to a particular orientation. These orientations are related to each other in significant ways. The central element in geometry is the line. What is the line's relation to the kingdoms of nature?

The geometry of the line always insists on a paired articulation, either static or active. This is reflected in nature. Although the mineral kingdom still puzzles me, there are two focal points on the lines of the spines of the kingdoms. In the plant kingdom they are vertical end points (canopy top and hypocotyl), and in the animal kingdom there is a horizontal tendency between head and rump.

Then there is the mystery of the principal symmetries of translation, reflection and rotation. How are they distributed among the kingdoms? It seems that all three symmetries are present in the mineral kingdom. In the plant kingdom, translation disappears leaving reflection (leaves) and rotation (in the branch nodes). Then in the animal kingdom, translation and rotation have vanished, leaving reflection.

The general orientation of a plant or tree tends to be vertical around the upright stem. In the animal kingdom, generally speaking the orientation of the spine is horizontal. Even in those creatures that might appear to tend to the vertical, like the penguin, kangaroo or ape, closer observation shows the horizontal dominating when the penguin swims, the kangaroo hops, or the ape runs on all fours. The human being, however, shows the vertical orientation again in the upright stance. This again shows the human kingdom as distinct from the animal.

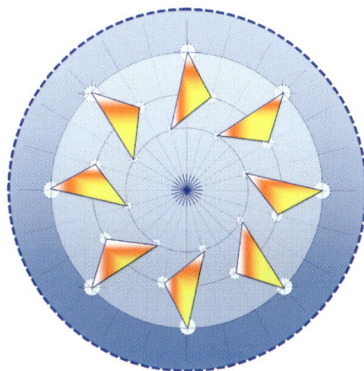

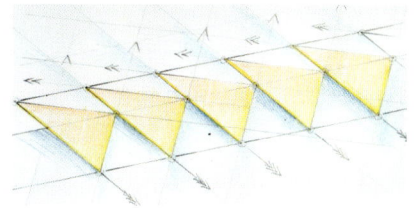

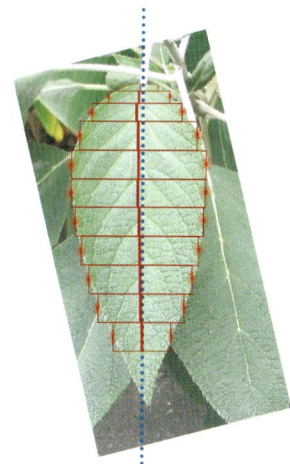

*Figure 1.19 Australian king parrot*
*Figure 1.20 Tiger at the Western Plains Zoo, New South Wales*
*Figure 1.21 Translation symmetry leaves the object in the same shape and orientation, but moves it in a certain direction*
*Figure 1.22 Reflection (or bilateral) symmetry is the mirroring of an object*

*Left:*
*Figure 1.23 Rotation symmetry leaves the object with the same shape but turned by a certain amount*

# 2. Desargues and Shadows

We saw in the previous chapter that the thought can somehow be in the machine – it wouldn't work otherwise – but do we see 'thought' or 'logic' in nature? What, for instance, is a shadow? The formation of such a thing should be simple enough to grasp. Can we see a *thought* structure here, a structure obeying logical ideas and rules? Is there a law that can describe for us how we might plot the shadows of wild olive seeds?

While it is possible to show transformations relative to Cartesian *x*- and *y*-axes, if we take a more general starting point, the thing gets particularly interesting. It is useful to begin with the Desargues' Triangle Theorem or, as it is alternatively called, the Shadow Theorem.

## 2.1 Desargues' Triangle Theorem

This theorem states that if two triangles are such that the lines common to corresponding points meet in a point, then the points common to corresponding lines meet in a line.

It is a lot easier to understand this statement by looking at a drawing of it. We see a radiating point marked *S* and a line marked *h*. Place a triangle between point *S* and line *h*. Three lines are put in point *S* so that they go through the three corners of the triangle. The sides of this triangle determine three points on line *h*. It is not hard to see how to draw the lower triangle, given one point on one of the three lines. It is good to try this exercise and it tests our drawing accuracy. Try it and you will see why.

The construction is straightforward and can start with a triangle in any three lines through the uppermost point (which is a proxy for the sun, hence *S*). The bottom left to top right line, *h*, in white represents the 'horizon'. The lower triangle represents the shadow of the triangular form above the horizon.

*Figure 2.1 Shadows of wild olive seeds.*

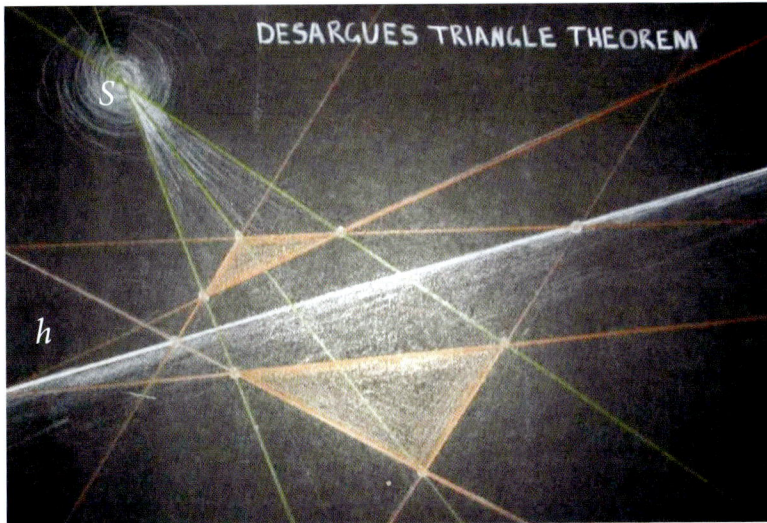

*Figure 2.2 Desargues' Triangle Theorem*
*[Harry Catterns]*

Another form of the same theorem is shown in Figure 2.3 except that the initial triangle *surrounds* the initial point, S. This drawing illustrates not just two triangles, but a whole family of them welling away from the initial point, and very quickly becoming somewhat strange; but they are still projective triangles. The larger triangles appear in two parts but projectively both the top and the bottom 'parts' are really one triangle, stretching through infinity.

It may be helpful to see how this works for a specific arrangement where the perspective of the construction is realised with a light source, an object, its shadow, and a line representing the horizon (Figure 2.4). The light rays out from the bulb (not exactly a point source, but it will have to do). The object, a glass tetrahedron, held above the surface casts a shadow on the level table. We can plot ray lines indicating the light path. The object projects a dark patch, the shadow that is changed in form and size though related.

This example shows that there is a clear underlying law in the phenomena of the shadow. It applies to all shadows on earth and beyond. It might seem to be a trivial example perhaps, though in fact it is not trivial at all. It has predictive as well as descriptive power, even if it is simple.

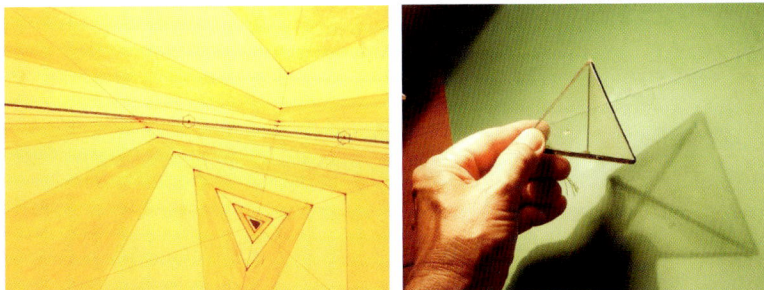

*Figure 2.3 Another form of Desargues' Triangle Theorem where the initial triangle surrounds the initial point*
*Figure 2.4 Tetrahedron shadow*

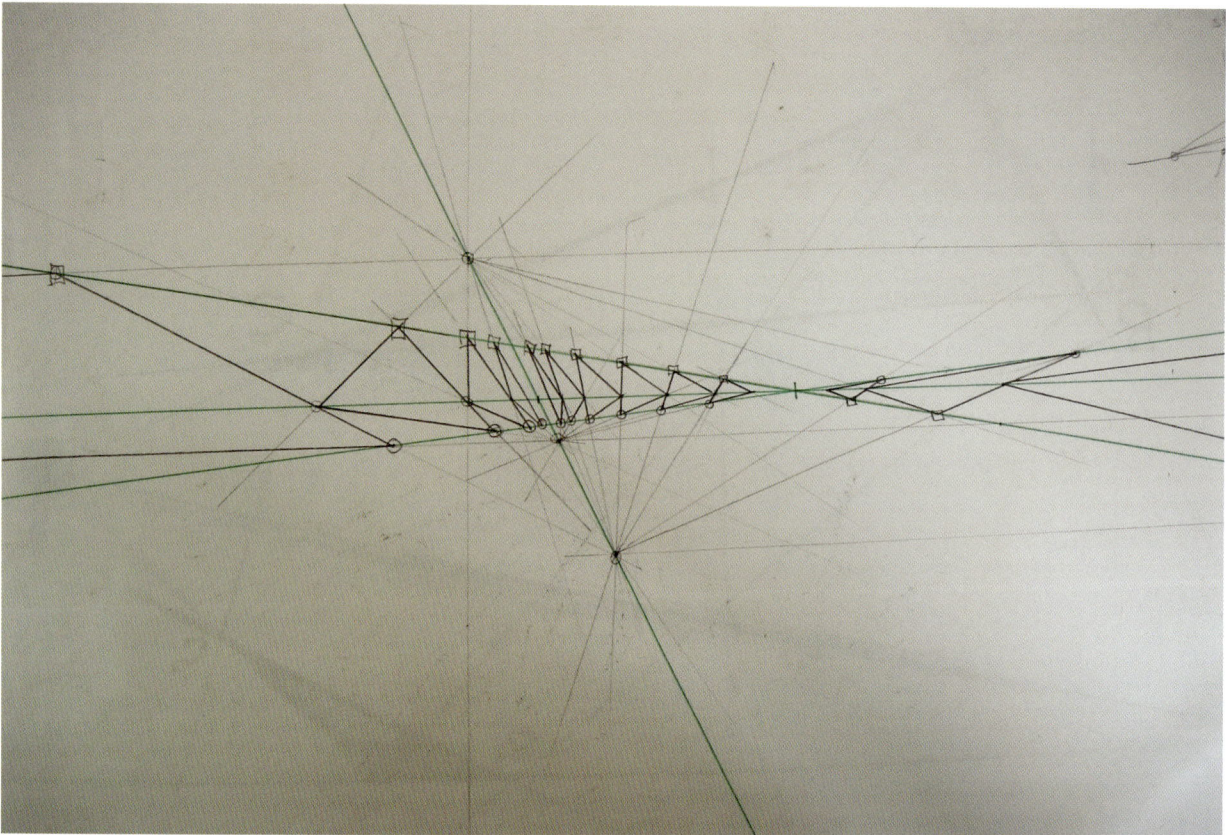

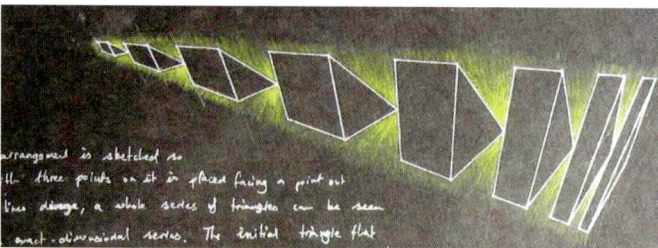

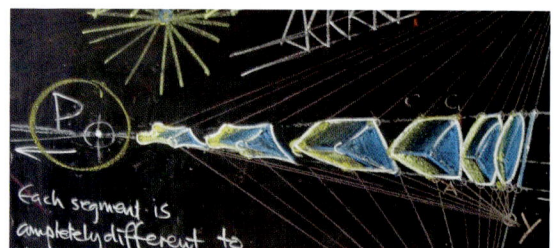

*Figure 2.5 A family of transforming triangles*

*Figure 2.6 'Vertebrae' drawing [Marco, a year 11 student]*

*Figure 2.7 Modification of the prisms of the 'vertebra' drawing*

## 2.2 Triangle series

Desargues' construction can also give us a whole series of related triangles. In Figure 2.5 we start with three lines through a common point, *S*, and another line, *h*. Draw any triangle within the three lines between point *S* and line *h*. Next draw a neighbouring triangle within the three lines such that their extended lines meet in the same three points of line *h* as those of the original triangle. Similarly, add more neighbouring triangles. Some of these head towards the point *S* as if attempting to vanish into it, others head towards the line *h* as if trying to melt into it. There are two opposite tendencies here. The triangles in Figure 2.5 have been extended both beyond (to the left of) line *h,* as well as beyond (to the right of) point *S.*

A rendering of this construction (Figure 2.6) can give the appearance of a series of solids, or triangular prisms. It must be noted that *none* of these little prisms are alike to each other. Their sizes differ, their angles differ, their orientations differ – yet they are patently parts of one family. When a fellow teacher saw this sketch he said, 'That looks like vertebrae!' so I have called it that ever since.

Perhaps there is something to this comment. Figure 2.7 suggests a little qualitative modifying of each prism, the rationale being that there is a different sculptural emphasis towards the point and towards the line.

Take any animal skeleton and observe both likeness and difference in each vertebra form. With the porpoise skeleton from the Pettigrew Museum in Scotland, many of the bony vertebrae seem only to change a little bit from one to the other. Do they all belong to the same transformation? Perhaps the famous Scottish biologist, Sir D'Arcy Wentworth Thompson's ideas on invariance need revisiting in the light of what Desargues can suggest for a simple family of triangles.

A vertebra is not a triangle, quite obviously, but they are both *forms.* And any form can *trans*form even if it is a bit more complicated – as is a vertebra. But is there an order to it? Is there some kind of field in which the separate forms are embedded? What gives the whole its integrated totality? We can see it clearly in the transformation of triangles, but what of vertebrae? We are a long way from fully understanding this, but the changing triangles suggested to me that there may be some ordering principle.

*Figure 2.8 Skeleton of a porpoise*
*[Pettigrew Museum, St Andrews, Scotland]*

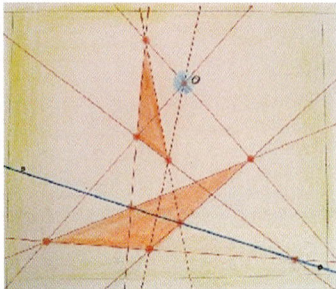

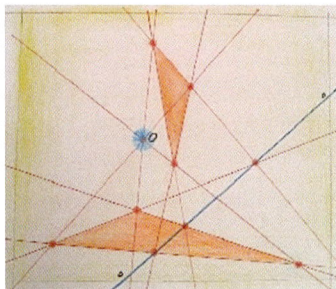

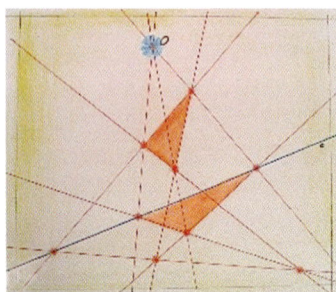

Figures 2.9–2.12 Variations of the Desargues construction

## 2.3 Variations and special cases

It is amazing that with *one* set of lines of the Desargues construction there are *ten different* triangle pairs to be found. A. Renwick Sheen describes this nicely in *Geometry and the Imagination,* (pp. 206f). I show four arrangements in Figures 2.9 to 2.12. Earnest students can discover the other six for themselves.

This construction, so important for shadows and perspective generally, has other implications. This is because the original configuration shown in Figure 2.2 can be arranged in various ways. If the initial point and initial line are given special locations, then the way the triangles are related become quite special. It is these special cases that nature seems to take an interest in and utilise in her own wonderful ways.

What we call Desargues' Triangle Theorem is, in a way, grossly oversimplifying a more general case. Any profile or two-dimensional form can be transformed. In Figure 2.13 a circle is transformed through a number of triangles (only one is shown) so that the shadow of the whole can be plotted. The circle transforms to an ellipse. This is similar to the shadows for olive seeds in Figure 2.1.

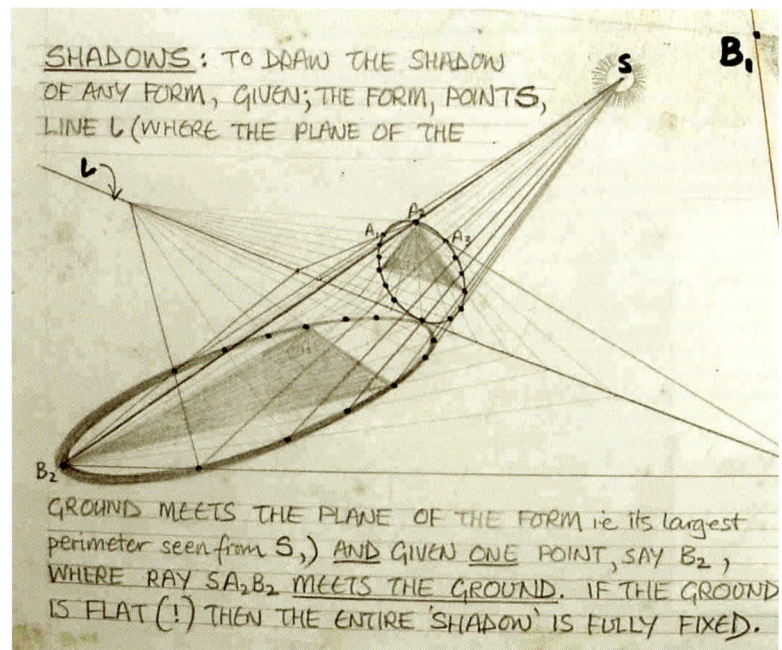

Figure 2.13 Shadows of a general form

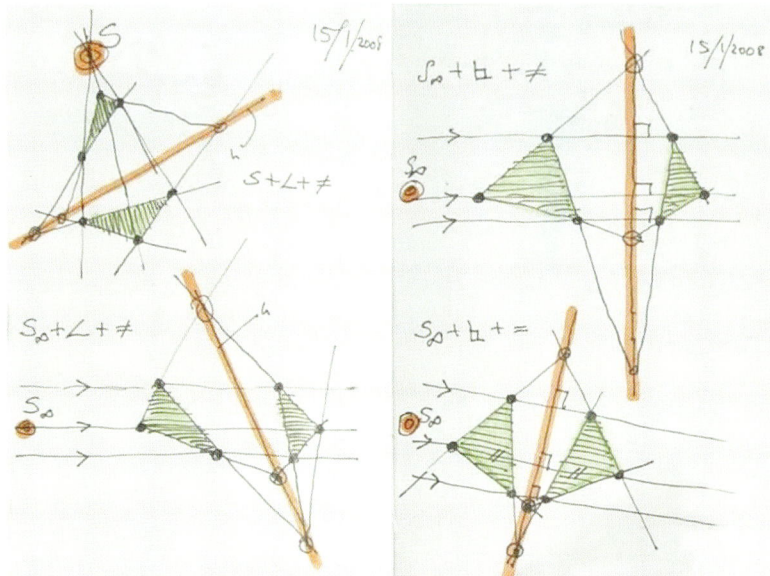

Figure 2.14 Top left: General Desargues theorem. Bottom left: Point S moved to infinity. Top right: Point S at infinity and line h perpendicular to the three parallel lines. Bottom right: Point S at infinity, line h perpendicular, and the triangle points are equidistant to the line h

## 2.4 Bilateral symmetry

The case with a random real line and a random real point in a plane has been covered above. But what if a few special positions for these fixed elements are considered? If we keep the line $h$ local (on the page, so to speak) but put the point $S$ very far away, in fact at infinity, then the three lines from it appear to be parallel (the three points can remain on the initial line $h$). Three stages towards bilateral symmetry are sketched in Figure 2.14.

First the point $S$ is put at infinity and the three lines become *parallel,* then the three lines are made *perpendicular* to line $h$, and then finally the distances of the triangle points to the line $h$ are made *equal.* We then have reflection, or bilateral symmetry, where the two triangles are reflections of each other. Thus bilateral symmetry is a particular case of the general Desargues construction.

Does this arise in nature? Yes: crystal facets often have it, plant leaves, orchids, butterflies, the animal world and human beings all show this symmetry. Typical leaves have a basic bilateral symmetry about their central vein when viewed from above (Figure 2.15). Orchids, *Phalaenopsis* for example, vividly show this kind of symmetry with a symmetry axis that is virtually vertical (Figure 2.16). The animal, viewed from the front, demonstrates this symmetry readily, again with an almost vertical axis of symmetry (Figure 2.17). And the human being is also a very strong expression of this form of symmetry.

This is one form of symmetry that arises as a result of gradually increasing constraints on the general Desargues configuration.

Figure 2.15 Bilateral symmetry in a variety of leaves
Figure 2.16 Orchids, Phalaenopsis
Figure 2.17 Wallaroo, a relative of the kangaroo

## 2.5 Translational symmetry

Another symmetry is translational symmetry. This appears to be the simplest of cases. Because (taking a triangle as an example) the triangular form does not change at all – the forms are all congruent, their angles, their side lengths and even their areas are the same, their orientation remains – they only change their position. Figure 2.18 shows triangles translating across the page.

What has to be customised from the original arrangement so as to achieve this? Once again point S is set at infinity so that the three lines through it again appear parallel, only this time the triangles have been placed between point S and line h. The three lines through S have been designated a, b and c and the three points in h have been labelled A, B and C. The third step is the most crucial and most interesting. The line h is now also moved to infinity (I use the practice, perhaps dubious, of a large dashed circle to symbolise this). If point S is at infinity, then S must be on line h. And points A, B and C must also be on this infinite line. What happens to the triangles? They are all congruent and identical – with corresponding sides and angles all equal – except for position.

This kind of symmetry suggests repetitions in the plane. Where do we see this in nature? On such situation could be at a micro level, presumably in the repetition of the basic atomic units. On a more macro scale this might suggest itself in crystal structure. Some crystals (like the galena in Figure 2.20) display a repetitious structure. They hint at a continuous repetition of rectangular, even macro square form elements. This structure

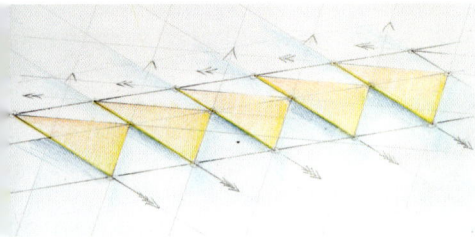

Figure 2.18 Translational symmetry

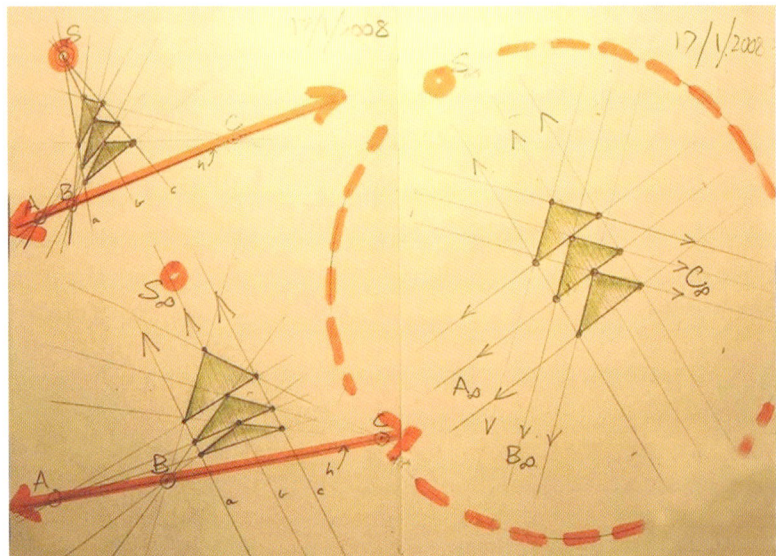

Figure 2.19 Top left shows the initial configuration. Bottom left has point S at infinity. Right shows line h also at infinity

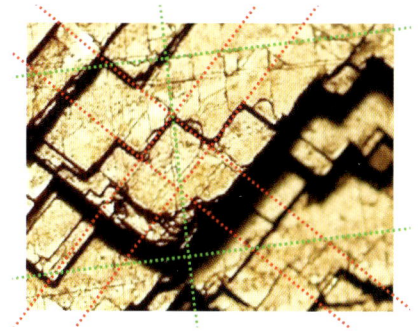

exists even in this chunk of galena: the lines are at 45 degrees to the apparent squares and rectangles in the photograph. This is clear in Figure 2.21 where a square element is picked out in red dotted lines and the 45 degree lines hinted at with green dotted lines. They go across both these perpendicular fracture lines. In Figure 2.22 there is a 'billiard ball' rendering of the molecular structure of a lead sulphide crystal. Clearly any surface would appear as a patchwork of squares.

Such tessellations can occur on still larger scales. I have seen rock shelves on the east coast of Australia where there is a relatively regular macro shaping in the surface, with eroded gaps almost like giant paving stones. Can we claim a translation of elements here too, just as in the Giant's Causeway in Ireland where there are predominantly hexagonal (six-sided) surfaces?

Nature builds with these repetitious forms and so do we. Think of the brick. Brick upon brick upon brick. We use this symmetry again and again in much building practice, both small and large. The windows of a skyscraper are often another good example of an architect's rigorous use of simple repetition.

*Figure 2.20 Galena*
*Figure 2.21 Galena crystal close up*

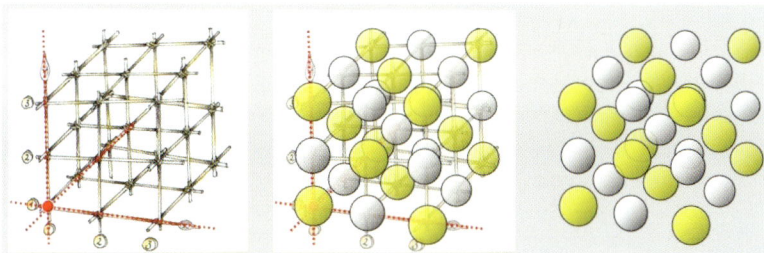

*Figure 2.22 Model of lead sulphide crystal*

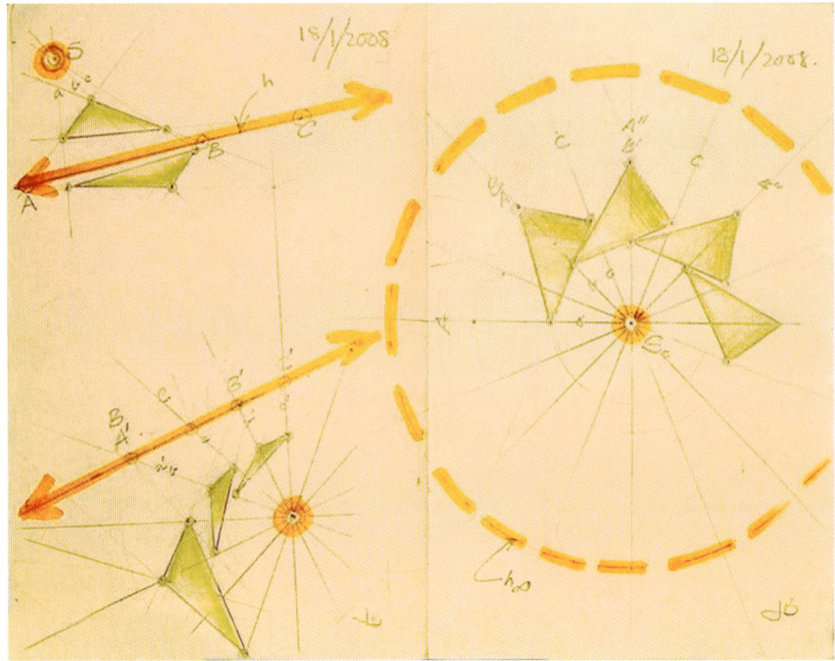

*Figure 2.23 General layout (top left). The three lines and three points move (bottom left). The line h goes to infinity (right)*

*Figure 2.24 Rotation symmetry in a flower*

*Figure 2.25 Rotational symmetry*

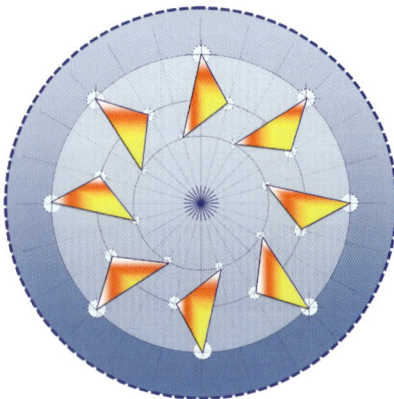

## 2.6 Rotational symmetry

There is one more very important symmetry to consider. This is where our initial point, $S$, is now given central stage. Line $h$ remains relatively local and the three lines $a$, $b$ and $c$ and the three points $A$, $B$ and $C$ come into motion. What does this mean? It means that the lines swing about point $S$ while, coupled with them, the points move along, or translate along, line $h$.

In Figure 2.23 the three initial triangles are shown again in the general layout for reference (top left). The diagram then shows how the triangles can move as the three lines, $a$, $b$ and $c$, swing and the three points, $A$, $B$ and $C$, translate (bottom left). They move in such a way as to make a passage around point $S$, gradually increasing then decreasing in size and altering in orientation and shape. Their path is an elliptical rotation. Any one point of a triangle moves in an ellipse with $S$ as a focus (so the three points would describe three nested ellipses). The next stage (on the right) is crucial. Place line $h$ at infinity (again shown as a large dashed circle to symbolise this special line), keeping $S$ central. The lines rotate around $S$ as before but the points translate now on the line at infinity. Surprisingly this gives the familiar picture of pure rotation. Hence we have a structure derived from the initial general case that delivers simple and exact rotation. The triangles become all the same and rotate about $S$ as centre.

And where do we find this fully regular circular rotation in nature? It is found everywhere, particularly in flowers, as in Figure 2.24. Here the role of the triangle is taken by each single petal. (Are they really the same, these petals? That is a question to follow up. And how much can exactness be expected anyway?) To draw the geometry of this, the line at infinity is simply assumed (Figure 2.25). The basic idea is that the structure needs both the central point and the peripheral line – even if we can never physically draw, see or get to it.

A rendering of the elliptical rotation inherent in the second stage of Figure 2.23 is given in Figure 2.26. In this sketch it's self-evident that each triangle is a part of a family. Each triangle is different in almost as many ways as is possible – yet they are still part of the same set. If one is drawn incorrectly it will usually stand out, for we seem to have an inner eye for the harmonious. How they appear depends upon where they are placed in relation to point $S$ and line $h$. In the sketch, the line $h$ is actually just off the page.

I had been curious to see if there was anything resembling this form of symmetry in nature. Then I came across a plant in a hotel leisure area on the north coast of Australia where this kind of symmetry *had* to apply. This is pictured in Figure 2.27 in a fan of leaves about their centre. There is an obvious bilateral symmetry in each leaf but also, around the tips of the leaves, this elliptical form or asymmetrical rotation, as I sometimes think of it. This does not seem to be common in plant species, but nevertheless exists in our world.

One interesting question that can be asked is, if the centre is apparent in this leaf fan, where is the line and what, if any, significance does it have? The answer is that it is part of the architecture, the geometry of the whole, even if we can give no immediate account for it. Why give any more credibility to the 'centre' than to the 'periphery'? From a geometric point of view the line, or periphery, cannot be ignored; it is as integral a part of the ideal construction as the centre.

I find it quite surprising to see that all of these symmetries are derived from the one basic design, and that there seem to be at least some representatives of all of these specialties in nature. As far as I know, the only place where the general situation arises is in shadows. The Shadow Theorem is therefore a fitting and alternate appellation for Desargues' Triangle Theorem.

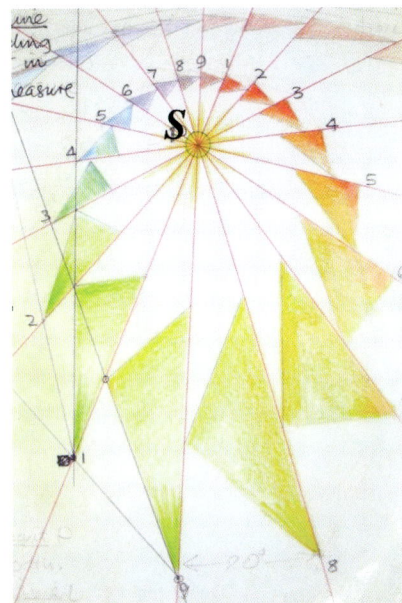

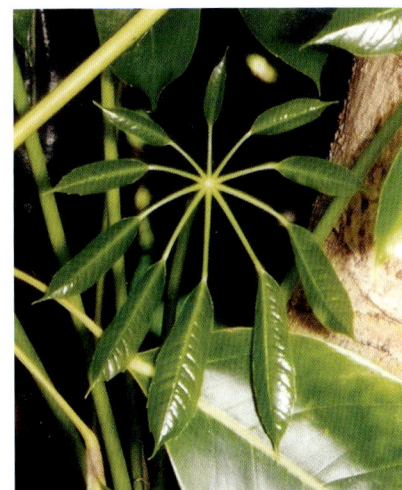

*Figure 2.26 Elliptical rotation*
*Figure 2.27 Asymmetrical rotation in the leaves of a plant*

## 2.7 Duality and polarity

Before we end this chapter, let's just touch on an aspect of geometry which will prove useful in later chapters. One of the fundamentals of projective geometry is a series of correspondences. For instance:

The joining of two different points is a line; and two different planes meet in a line.

The joining of a line and a point (which is not in that line) is a plane; a plane and a line (which is not in that plane) meet in a point.

In these two statements *point* and *plane* are interchangeable, while *lines* remain. In other words, what holds true for points and lines in a plane, holds true for planes and lines in a point.

This is known as the principle of *duality* or *polarity,* and each proposition is called the dual or polar of the other.

If, instead of working in space, we confine ourselves to the plane, the basic entities we work with are just point and line. We can interchange these two elements as the duals in a plane.

*Figure 2.28 We can imagine the ellipse in a plane (black lines joining the points) or as an elliptical cone in a point (green planes)*

For instance, three points define a triangle, and conversely, three lines define a triangle.

What is more difficult to imagine (mainly because we are not so familiar with it in our everyday life) is to confine ourselves to a point and all the lines and planes passing through it. This would be a geometry of the point. There we can state that:

Any two lines define a plane; and any two planes define
 a line.

And we could similarly dualise in this realm by interchanging plane and line.

So far, all our form considerations have been in the plane. But everything we can do in the plane we can do in a point. This should be clear from Figure 2.28 where we can imagine the ellipse in a plane and the elliptical cone in a point, intimately connected. In the plane we 'draw' with point and line. In the point we would – if we could – 'draw' with line and plane only. This is very hard to imagine, but we can allude to this in the drawing.

# 3. Geometric Elements and their Gestures

I want to pursue this thinking with a view to seeing our world, not only from a fragmentary, particle perspective – which is the reductionist malaise – but also from a line-like *and* a plane or surface outlook.

## 3.1 The planar element

I remember once a few years ago crossing the North Sea returning to England from Norway. The sea was virtually flat, a so-called millpond, except for our ferry's turbulent wake. This mirror of a

*Figure 3.1 The crystal surface of pyrites*

sea was surely far from common! Here was flatness – an almost uncanny flatness in a usually active surface. It was so flat that we could see the clear reflection of a gull swimming a number of metres away from our boat in what would normally be a choppy, swirling and frothy arena. Reflection relies on the reflecting surface being *flat*. Here is flatness, the planar in its essence.

The surfaces or faces we see on crystals are also often very flat, like pyrites, or the slightly rougher, yet basically flat, faces of the orthoclase feldspar crystals, or smoky quartz. It is clear that there are highly reflective facets making up the exposed structure of these crystals, so much so that they can be seen as many natural little mirrors. Here we see a part (or facet) of the infinite plane again.

To return to the geometry, how do we draw a plane? Any purely geometric plane is impossible to illustrate. We are reduced to picturing slivers of planes, little pieces of them cut out from the whole. Can our plane even have any density? Of what is our plane constituted? It can be thought of as dense with its elements. There are two such elements – lines and points, which are just as insubstantial as the plane itself! Every point has an infinity of lines in it. And every line has a whole infinity of points in it.

Does this also mean that even every facet 'contains' such infinities? Indeed it does. Such a facet is however bounded by these other two elements, line and point, or edge and vertex, junction and corner. Take a facet of the quartz crystal. It can be seen as bounded by several points and an equal number of implied lines.

What is the minimum number of points that can define a plane? It is *three* points. And what is the minimum number of lines that define a plane? It is *two* lines that meet (the latter is important, as the two lines could be skew, thus never touching). One line and one point (not in the line) will also define a plane.

The geometric or conceptual plane is an ideal of infinite extent, while the planes we see in the world around us are finite (and imperfect). And yet there must be a contiguity. It is as if, when we see a crystal's flat surface, the infinite, ideal plane arranges the numerous point-like molecules in the most marvellous order.

Perhaps this invisible, infinite plane determines the macro form of surfaces like the striations in a pyrites crystal (Figure 3.5), or the cleavage planes in fluorite or diamond that makes possible the relatively easy formation of an octahedron (Figure 3.6). (The coherent profile arrived at relies on a definite planar slicing being possible, as shown diagrammatically in Figure 3.7).

*Figure 3.2 Orthoclase feldspar [Albert Chapman Collection, Australian Museum, Sydney]*
*Figure 3.3 Smoky quartz*
*Figure 3.4 Quartz facets*

Can we in fact think of such surfaces as being both the material micro mineral and the infinite, invisible plane? Are they not two sides of the same coin?

Invisible surfaces are everywhere. The clouds often seem to ride at a particular level (that they are caused by pressure or temperature gradients is not the point). What we *see* is a suggestion of the planar element that carries the water vapours.

## 3.2 The linear element

The line is the middle of the three elements. Does this make it more significant in some way than the other two elements? The line has both an intensive aspect and an extensive aspect. Look along its length, and all we see is a point, its section. Look from elsewhere and we see an infinitely extensive entity. It has a foot in both camps, and can be thought of as the element in the middle between plane and point.

Where do we see some semblance of the line? We can never claim to see a whole line – at best we can say that what we see is a *line segment* or a line *interval*. Nature shows us a multiplicity of line segments in grasses, a growth of bamboo, or the fine petals of a wattle flower, or the spines of a sea urchin. In all this we really only see hints of a line, for lines are infinitely long.

Often what we call a line is only the meeting place of between two surfaces of different colour, as in Figure 3.10, a view in rural New South Wales: the sunlight contrasts with the grey cloud, with a clear swathe of light rays, the house roof ridge contrasts the background foliage, and sky, the tree trunk behind the house is against the sky, the lower cloud surface is in contrast to the background coloration, the many vertical posts, and the road marking contrasts the bitumen surface. This linear element weaves itself across our vision with both natural formations and man-made artefacts.

Again, where is the line around us? There is obviously much more in the perceptual world that suggests straightness. It is anywhere the *straight* emerges as the meeting place of two differently coloured surfaces. Or it is even in the meeting place of different densities in the

*From top:*
*Figure 3.5 Striations in pyrites*
*Figure 3.6 Cleaved fluorite*
*Figure 3.7 Octahedron built of cubes*
*Figure 3.8 Cloud layers*

*Figure 3.9 Ornamental bamboo*

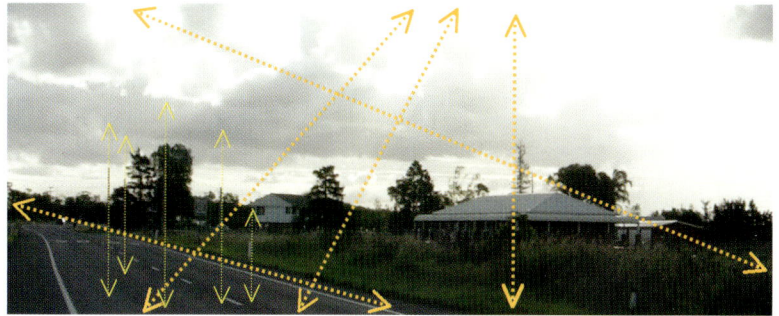

*Figure 3.10 Lines in a rural scene*

rocks, as in Figure 3.11. Or it is in the plant stems from agapanthus to grasses, from pine trees to gymea lily shaft, from thorns to the very spiky spines of such as the sea urchin, *Echinometra matthaei,* Figure 3.13.

In some places, such as a towering cityscape, we would have very little notion of perspective were it not for our invisible, continuous straight line.

Here we really do see lines, or rather just line segments, as representing something that is seen in many guisesacross a cityscape or landscape.

There are two distinct possible ways to define a line. Any two points in space will define a line – the line that passes through them and going beyond them (the two points, $P_1$ and $P_2$, define the line, $l_{12}$, as in Figure 3.15). And any two planes define a line (the two planes $\pi_1$ and $\pi_2$ define the line, $l_{12}$ as shown in Figure 3.16). Note that if the planes are parallel they 'meet' in the line at infinity. We can find an everyday illustration of the meeting of two planes in any room. Two walls meet in an edge from floor to ceiling.

*Figure 3.11 Rock boundaries*

*Figure 3.12 Thorns*

*Figure 3.13 Sea urchin, Echinometra matthaei [Ashley Miskelly]*

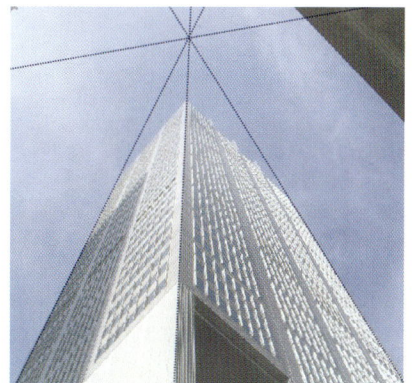

*Figure 3.14 Lines of vanishing perspective in a towering cityscape*

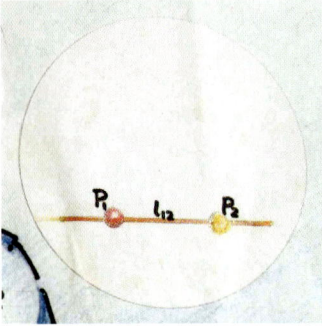

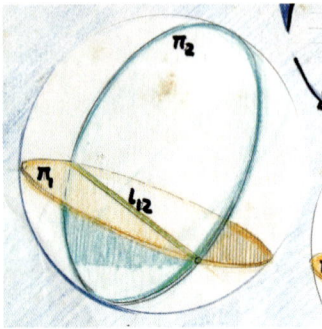

*Figure 3.15 Two points define the line*
*Figure 3.16 Two planes define a line*

We need to be mindful that these seemingly abstract lines are full to the brim: they contain an infinitude of points. One might even say that a line consists of the points lying along it. In nature these things can, needless to say, only be partially represented. But they are out there and it is worth illustrating this. Recently in Queensland, I saw many clusters of bamboo. They suggest *line*-like stems with sequences of *points* along them (Figure 3.17).

But it is equally valid to describe the line as containing an infinitude of *planes*. The planar element sweeps *around* the line while the point rides *along* it. The planes in a line are crudely illustrated in Figure 3.18. Where might we find an image of the planar in a line in nature? Two instances come to mind. One is the two planes that come together to create an edge in any crystal. The other is in a little pine tree, used ornamentally, and known colloquially as the 'book pine' (Figures 3.19 and 3.20). Each leaf – if that's what we can call them – appears almost flat and vertical.

Both these aspects – rotation of planes and translation of points – need to be considered at the same time. This will be explored later in the book when the combined activity of the nodes along the line are related to the planar working around the line together (Chapter 13), as it is particularly relevant to the plant world.

*Figure 3.17 Bamboo and nodes*

## 3.3 *The element of the point*

The point is our culture's favourite element. We think that we can explain everything using its components. The little bits, the cells, the molecules, the atoms or even subatomic particles are called upon to provide answers about the *whole* of which they are merely *parts.* There is something absurd about this reductionism. The smaller we go, the less we see of what it is a part. As the whole is reduced to parts, the context simply vanishes.

This viewpoint – a prejudice if seen as *exclusive* – has dominated western thought for centuries, and has led to extraordinary discoveries despite being one-sided. (However, the notion that the whole is determined by the bits is slowly being questioned, as for instance in Denis Noble's *The Music of Life).*

A point as such has no magnitude, so it is meaningless to ask how much bigger something is than a point. But we can see that a point suggests *intensification* rather than an extension. This can be applied from the smallest imaginable particle to the largest star, for the largest star is a virtual nothing in the vastness of the cosmos we believe to be out there.

Just as a plane can readily be defined by any three points, so the reciprocal is true: a point can be defined by any three planes. In Figure 3.21 the three planes $\pi_1$, $\pi_2$ and $\pi_3$ cut each other in pairs of lines ($l_{12}$, $l_{13}$, and $l_{23}$) which all go through the single point $P$.

These two definitions are *polar* to each other. This polarity makes sense if we do not give either the point or plane any priority

*Figure 3.18 A fan of planes through a line*

*Figure 3.19 Book pine*
*Figure 3.20 Leaf close-up*

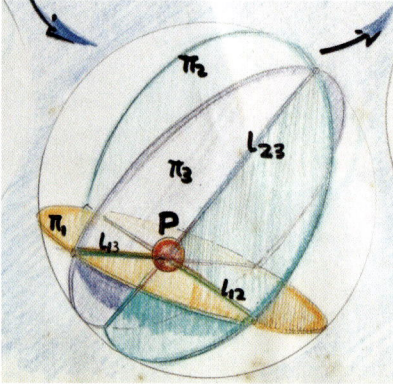

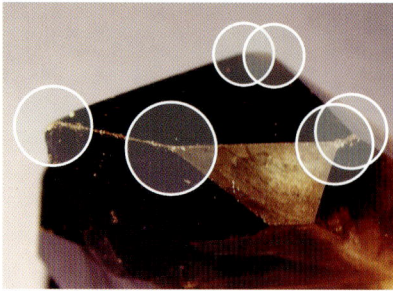

Figure 3.21 *Three planes determine a point*
Figure 3.22 *Vertices (points) are formed where three facets of a crystal meet*
Figure 3.23 *Waratah seeds*

over the other. So the definitions are completely reciprocal, despite the reciprocal elements being utterly different to each other in quality. Yet their mutuality is patently clear. Might this suggest that the pointwise approach to nature, that has been so successful, could be complemented by a planewise approach?

We see an example of the definition if look at three facets of a quartz crystal. Each adjacent face meets in an edge. Where the three edges coincide there is a point, or a *vertex*, as it is also called. Some of these vertices are circled in Figure 3.22.

In the plant world there is a tendency towards the point-like in the seed. We sometimes talk of the seed-point when there is a beginning to things. For instance the Waratah seed is only slightly larger than a match head. Yet we would still see it as a point compared to what it can become, for in suitable conditions it grows into one of the largest flowers in New South Wales, and is used as emblem of that state.

The abstract point contains far more than at first appears. While just two lines are sufficient to determine it, each point contains an infinitude of lines passing through it. And it also contains an infinitude of planes passing through it. Perhaps the seed also contains far, far more than we might at first think.

## 3.4 *The interdependence of the elements*

We cannot really think of any of the three elements on their own, for each implies the others. In any crystal form this is self-evident, as when we have a corner or vertex there are at least three lines

Figure 3.24 *Waratah flower*

meeting it. And each of these lines is the intersection of two planes. The *faces* of a constructed Platonic solid act as a modelled example – these pentagonal faces connect in three *lines*, which meet in one *point*, and this occurs twenty times (Figure 3.25). The converse of this would be where three points, creating three lines, bring about a single triangular plane, in this case, one of the twenty faces of the icosahedron (Figure 3.26).

The three elements always work together. They are considered self-evident axioms – *a priori* – and are intuitively obvious to most of us, except perhaps, as Lawrence Edwards used to say, 'to a roomful of mathematicians' (*The Vortex of Life,* p. 18).

Any of the static forms so far described can be represented from the point of view of one or other of the three basic elements. For example, it is possible to sketch a tetrahedron from the three aspects. Figure 3.27 shows the 'billiard ball' or point-wise approach (top left), the 'stick' or line-wise approach (top right), and the planar or plane-wise approach. The last is the most common presentation as it lends itself to cardboard models.

Figure 3.28 summarises the separate characterisations of the elements we have examined. The bottom left and top right determine the line, while top left and middle determine the plane (from line and point) and bottom middle and right determine the point (from plane and line).

Something that shows the interaction of the three geometric elements in nature is the notion of anastomosis. Anastomosis concerns how something covers a surface or fills a volume, as a

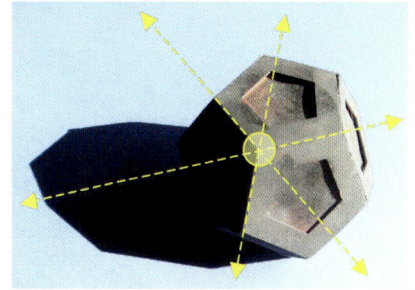

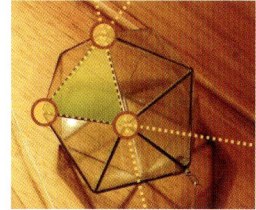

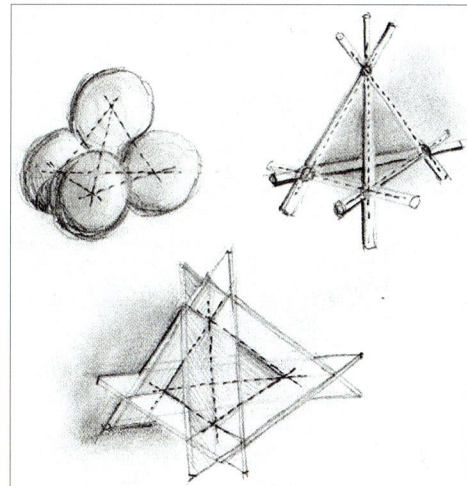

Figure 3.25 The corners of a dodecahedron
Figure 3.26 Faces of an icosahedron [Christel Post]
Figure 3.27 Point, line and plane tetrahedra

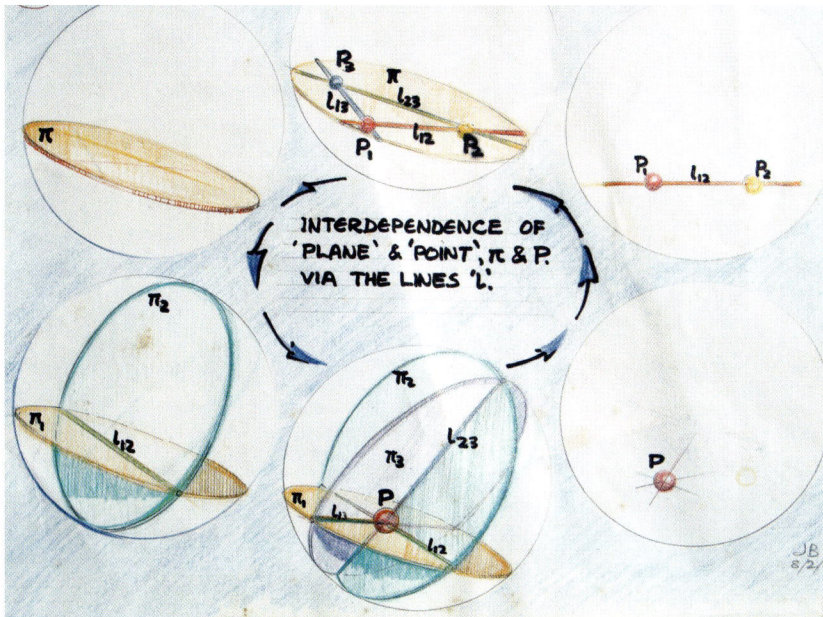

Figure 3.28 The mutual relationship of the elements

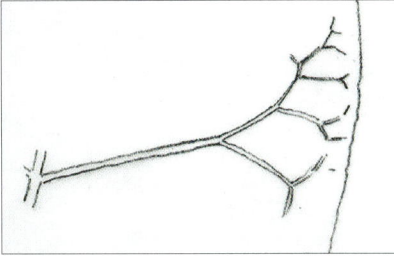

*Figure 3.29 Branching*
*Figure 3.30 Leaf veins and connectivities*
*Figure 3.31 Branching of a dragon tree*

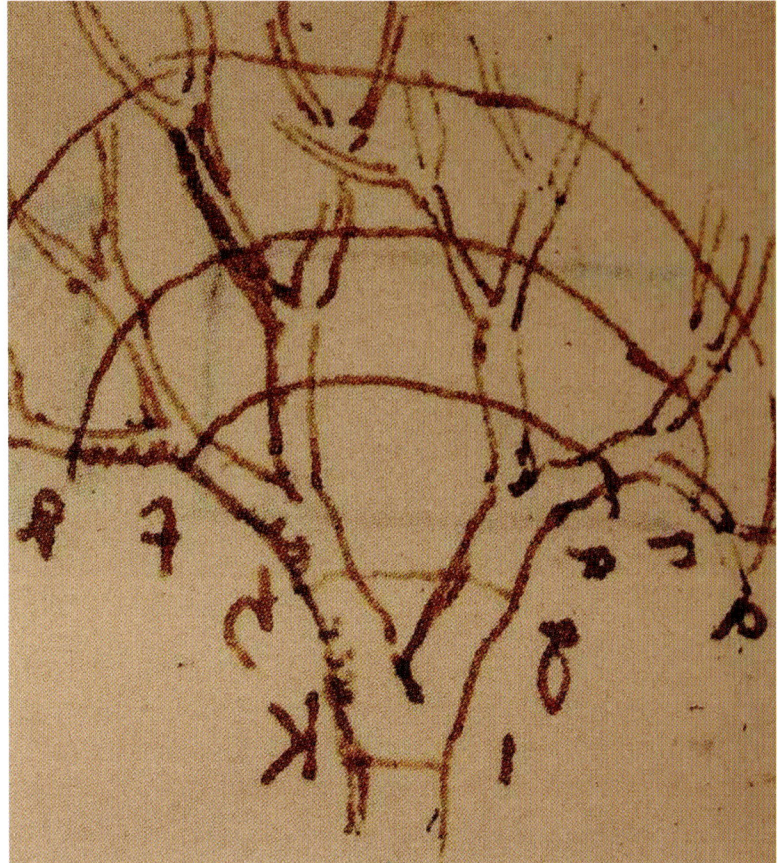

*Figure 3.32 Sketch by Leonardo*

network of interrelated connectivities. For example, the veinage of a leaf often has a particular character. From the junction with the stem through to the periphery of the leaf blade there is a very interesting and frequently represented transition. This is a case where point and line work together (in the plane, for leaves are basically planar). At the nodes there is a branching into line segments, then again, further branching at the next nodes, more branching, and so on, getting finer and finer (Figure 3.29). Compare this to the veins of a leaf (Figure 3.30) or to the branching of a dragon tree (Figure 3.31).

To me this suggests a transition from one element to the other, from junction to curve, node to periphery, from point to line. Leonardo da Vinci noticed this; his drawing could be the veins in an organ, a tree or a leaf (Figure 3.32). These patterns, of course, looked like the bifurcations which result from the logistic equations in so-called chaos theory as popularised by Robert May and

*Figure 3.33 Logistic branching*

others in the 1970s. A visualisation of a simple logistic equation has a similarity to these images (Figure 3.33).

It was only when alternatives to Euclid's fifth postulate (concerning parallel lines) were seriously explored in the early nineteenth century that whole other geometries arose. Nikolai Ivanovich Lobachevsky, a Russian, and independently Janos Bolyai, a Hungarian, discovered what was later called non-Euclidean geometry. This led to the development of synthetic or projective geometry which regards points at infinity as equal and transferable to those in Euclidean space.

# 4. Symmetry in Nature

*Figure 4.1 Mountain devil fruit*
*Figure 4.2 Butterfly symmetry*

## 4.1 Bilateral symmetry in plants

There are many things that show this kind of symmetry. It is to be found in the human being, the animal, the plant and in the mineral. It is perhaps the most common form of symmetry. Other symmetries are less evident, and sometimes not at all.

Figure 4.1 is the fruiting form of a native Australian plant known as the mountain devil. It is an excellent example of bilateral symmetry at work. Note that the reflection is about a central line. However, it is not only in the picture plane – it is in space. The beautiful butterfly (Figure 4.2) confirms this for, while the wings do indeed appear flat, the symmetry we speak of is centred about the central vertical plane of the insect

body. For the line of symmetry, viewed from above, is really a plane extending above and below, and the body itself is a three dimensional entity.

This symmetry plane can be found in the form of many insects and animals, as well as in the human being. The vast majority of creatures have a bisymmetric body plan, and one has to actively look for the few exceptions. This symmetry is true *externally* in both human and the animal. But inside the skin it is a different story. In both animal and humans, the organs are not mirrored.

In the plant world, the leaf shows this symmetry clearly, as in Figures 4.3 to 4.8. But take care. Figure 4.9, an Australian native, certainly looks right, but is this bilaterality exact, or a mere tendency? Figure 4.10 indicates the latter but it is quite close. The yellow dots/stars in this figure are nearly the same distance from the central vein (the reference rectangles are identical) as each other. Note that some of the spikes are not exactly replicated on each side of the central spine. For me there is enough to suggest that there is some kind of formative force that tends

*Below:*
*Figures 4.3–4.8 Various leaves*

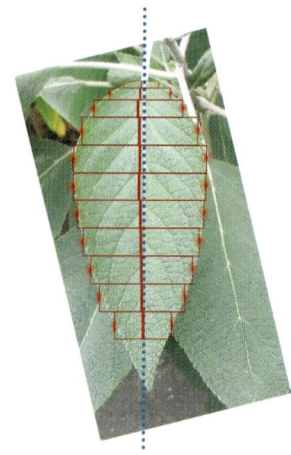

Figure 4.9 *This looks close to bilaterality*
Figure 4.10 *Geometry indicates it is quite close*
Figure 4.11 *Bilateral leaf test*

Figure 4.12 *Reflective test*
Figure 4.13 *An asymmetrical begonia leaf* (Begonia manicata)

towards ideal bilaterality, but does not necessarily achieve it. I think of tendencies towards rather than metric precision. Though why one side *should* mirror the other still remains a mystery.

In Figure 4.11, the bilateral leaf test, I choose a leaf that it would be easy to assume is fully mirror reflection about the central vein. How close is it? I have marked left and right leaf edges at various points, and marked the midpoint between them (thick red line). The blue dots show the central rib of the leaf. The differences are apparent, but it is not an unreasonable expression of the bilateral, being truer towards the leaf tip. Note too that the vein pattern on either side is similar but not the same. This similarity without strict identity is intriguing. The two sides are out of step but still close. This is common.

One might think that the leaf in Figure 4.12 is a good candidate. Visually it is, but it is not exact. This time I have overlaid a blue many-sided polygon on the right-hand side of the real form, assumed a central rib (blue dots) and flipped this polygon form about the central axis. This gives the yellow polygon. It is now easy to see how close (or otherwise) the left-hand side is to the right. Also inserted are the red parallel lines through the bilaterally symmetric points, and perpendicular to the central line, which imply a theoretical point at infinity.

Once again I would assert that there is, in this leaf organism, a strong tendency to realise this bilateral form of symmetry. Also recall that this involves elements at infinity so the conceptual context in which the leaf exists is infinite.

Not all leaves have this tendency to bilateral symmetry. Some have a marked asymmetry, and this geometric paradigm has yet to be explored. The large begonia family is rich with this asymmetry; *Begonia manicata* is shown in Figure 4.13. The question can still be asked, is there some systematic, valid geometry to this asymmetry? I cannot answer that at present, though I suspect it will involve harmonic constructions (see Chapter 14). While such a leaf may be asymmetric in itself, it may be bilaterally symmetric to its counterpart on the other side of the stem, as in Figure 4.14.

There are also very many plants with bilateral flowers. These include the pea family (*Leguminocae)* and the many hundreds of orchids (Figures 4.15 and 4.16). There is no doubt of this bilaterality.

I tried a superimposition technique to assess how true this bilateral flower symmetry is. First set the original image around an estimated vertical centre line (Figure 4.17). Half the image is copied, flipped horizontally and superimposed on the original. For clarity the flipped copy is black and white with about 50% transparency (Figure 4.18). With the chosen orchid the superimposition is not perfect, but reasonably good, and enough to see a striving towards bilaterality.

This symmetry is not lacking in the other three kingdoms. A few examples follow here, just to demonstrate the ubiquity of this kind of symmetry. If we choose to take geometry seriously, and not simply as an arbitrary plaything, then all the kingdoms have to be thought of as part of vast contexts.

*Figure 4.14 Asymmetric but bilaterally symmetric to the leaf on the other side of the stem; Figure 4.15 and 4.16 Orchids; Figure 4.17 Symmetrical orchid; Figure 4.18 The left side of the orchid has been copied, flipped and superimposed in black and white on the original*

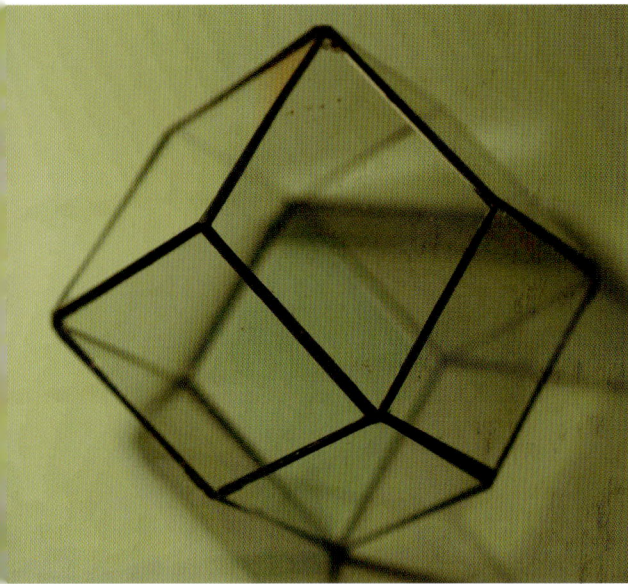

*Figure 4.19 A glass rhombic dodecahedron [Christel Post]*
*Figure 4.20 Garnet*

## 4.2 Bilateral symmetry in minerals

The crystal symmetries of the mineral world are well known together with their properties and structures. Today's natural physical science has a powerful consciousness of the material, so I will only show a few examples here.

One geometric form that is close to a familiar crystal shape is the rhombic dodecahedron.* A glass model is shown in Figure 4.19. The familiar crystal is the garnet, a small deep-red piece that has facets in the shape of a rhombus (or very close to it). Figure 4.20 is a small garnet crystal that I purchased at a little mining town in the outback of Australia. It gives an excellent bilateral expression. This crystal form is very close to a perfect rhombic dodecahedron.

Such an ideal form has twelve faces, each formed as a rhombus with particular proportions. A rhombus is a transformed square where the diagonals are not equal. In this case the rhombic dodecahedron structure determines the proportions and dictates (through its relation to a cube) that the diagonals are in the ratio $1 : \sqrt{2}$, or approximately $1 : 1.414$.

To determined this ratio, we can look end-on to a rhombic dodecahedron and see four of its faces covering a cube (Figure 4.21). If the cube is of side length 2 units, then (by symmetry)

---

* A dodecahedron is a regular body of twelve five-sided faces, sometimes also called a pentagon-dodecahedron. The rhombic dodecahedron is a semi-regular or Archimedean solid with twelve rhombic faces. A rhombus has four sides of equal length, but not square (a diamond is a rhombus).

Figure 4.21 Rhombic dodecahedron and cube
Figure 4.22 Determining the angle of the rhombus
Figure 4.23 Double symmetry

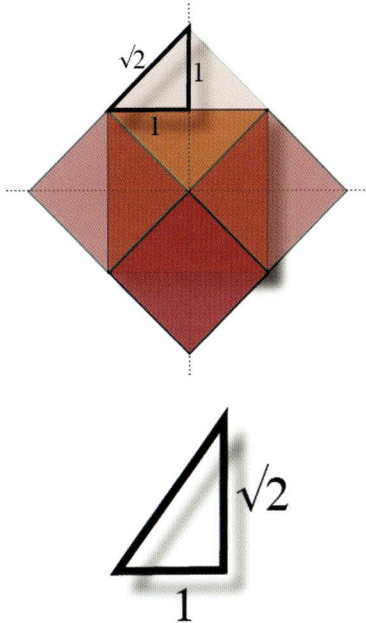

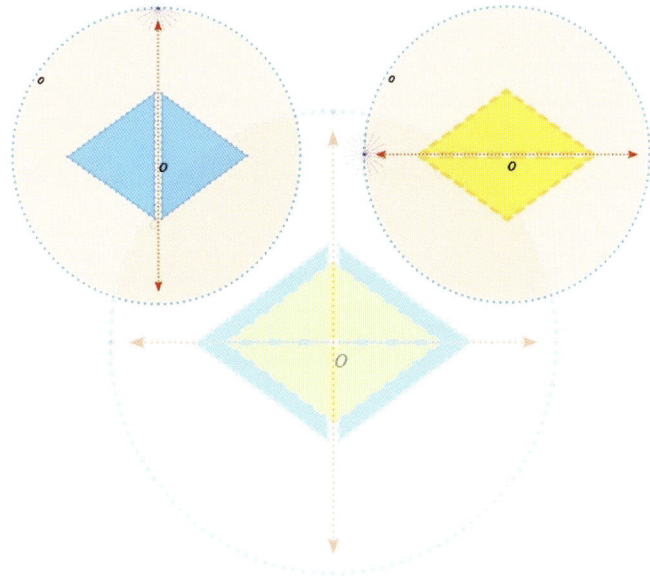

the right-angle triangle above the cube has its equal sides of one unit each. Hence with Pythagoras' Theorem $(a^2 + b^2 = c^2)$ we can easily calculate the longest side (the hypotenuse) of the right-angle triangle, as $\sqrt{2}$ or about 1.414. The parallel sides of the facet are at a specific angle to the horizontal diagonal. This angle, $\Theta$ *(theta)* is readily calculated from Figure 4.22: $\Theta = \tan^{-1}(\sqrt{2}/1) = 54.7356°$ (about 54.7°).

Such a crystal has a similar bilaterality to leaf forms, but with one significant exception. This mineral is bilaterally symmetric about both its diagonal axes, and in fact through both diagonal axes of each face. The mineral permits itself multiple expressions, in the same shape, of this kind of symmetry form. This multiple symmetry is rare in the plant world, hardly seen at all, although there are some fourfold flowers and fourfold stems showing a double symmetry.

Even the simple rhombus (and thus also the garnet) demands the infinite elements, as the geometry necessitates that there are two points perpendicular to each other on the line at infinity (Figure 4.23).

I use the garnet as representative of all such crystals with such regular facets. Pyrites (Figure 4.24) illustrates a rectangular double symmetry. The rectangle also requires two points on the line at infinity to define the parallels. We can see this in Figure 4.25 where points *A* and *C* have to be on *l*, the line at infinity. The diagonals *B* and *D* are also on this line.

Figure 4.24 Pyrites rectangles

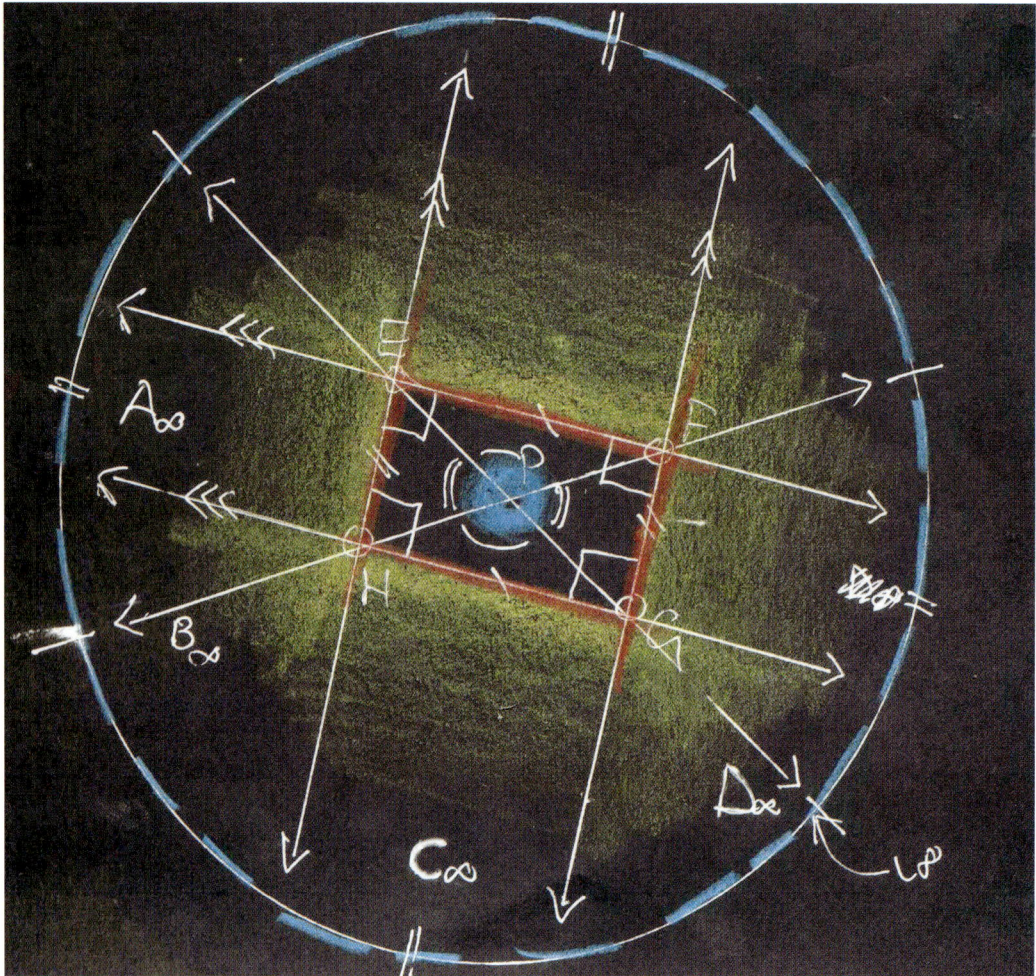

In orthoclase feldspar (Figure 4.26) there is a double symmetry in space rather than in the plane only. As with the plant leaf there is some inaccuracy. In this I can see the natural striving towards the ideal. To demonstrate how the garnet seeks to maintain a parallelism on both axes, at least, and yet doesn't find the precise rhombus, I have included another garnet (Figure 4.27) where we have a parallelogram instead. It is as if the garnet is 'looking for' the rhombus.

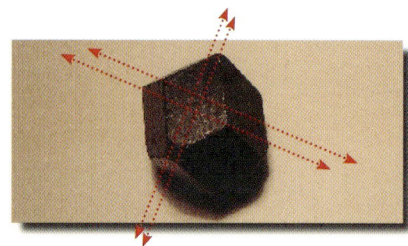

Figure 4.25 Rectangle in context
Figure 4.26 Orthoclase feldspar
Figure 4.27 Non rhombic garnet

## 4.3 Bilateral symmetry in the animal and human kingdom

Animals also display a strong bilateral symmetry from one aspect: the left-right symmetry seen from the front or from the back (although this latter should not be over-emphasised). The basic structure of vertebrates is that a vertical plane divides left side from the right (this does not apply to internal organ arrangements, which are hardly symmetric at all).

But how precise is this symmetry in animals? Short of detailed measurement from an estimated centre line, or plane, it is not easy to ascertain. However the visual method used for the orchid flower may help. Here are a couple of examples. I took a picture of a brush-tailed possum (which had got stuck in our attic during some roof repairs). The picture should be as near to a true front view as possible (Figure 4.28). Estimate the centre plane (a line as viewed from the front). If necessary rotate the image to make the centre line vertical. Flip horizontally about this centre line (Figure 4.29). Superimpose the copy over the original image, lining up the centre lines of the two images, and making the overlapping copy 50% transparent (Figure 4.30). Note the correspondence of the two images. This is a good match – you would hardly know you are looking at two images.

*Figure 4.28 Brush tailed possum showing bilateral (left-right) symmetry.*
*Figure 4.29 Possum image flipped*
*Figure 4.30 Possum images superimposed*

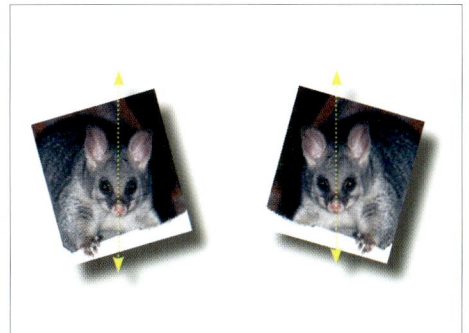

Some interesting Australian animals (except the tortoise) are seen in Figures 4.31 to 4.34. Another superimposition is of a former family pet, Suzie. The methodology is the same. Here I'll only to comment that the two images are very close – except for Suzie's trademark one ear up, the other down (Figures 4.35 and 4.36). I include a couple more native Australian animals for reinforcement – a koala and a wallaroo. There is some asymmetry in these apparently bilateral forms, but from the given examples it does not appear to be much.

*Figure 4.31 Cassowary*
*Figure 4.32 Platypus*
*Figure 4.33 Echidna*
*Figure 4.34 Tortoise*

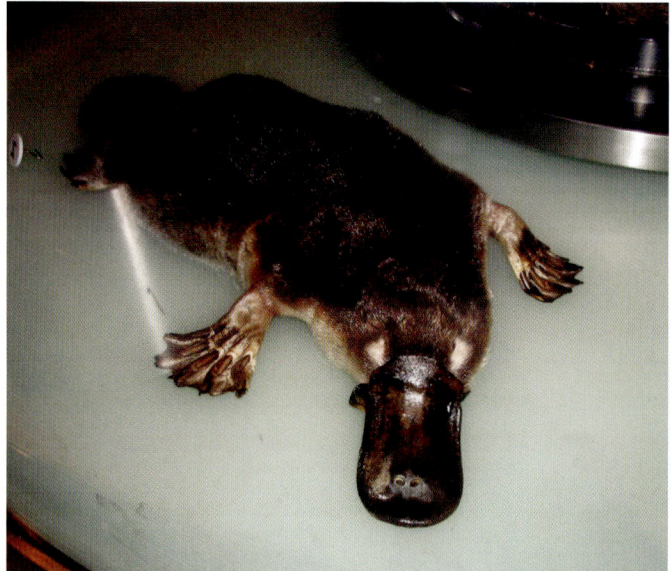

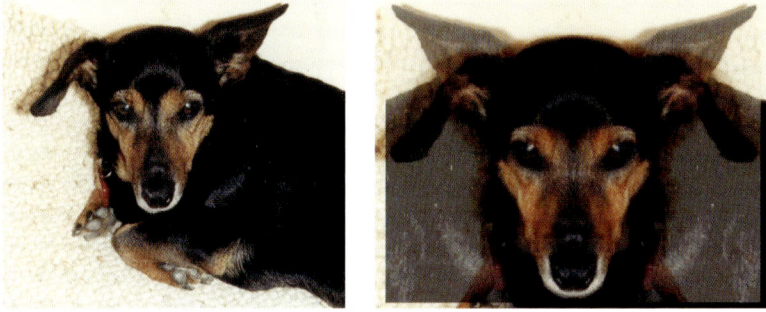

*Figure 4.35 Suzie, the dachshund*
*Figure 4.36 Suzie, superimposed*

This is a cumbersome procedure for stating the obvious. The eye can *see* that there is patently this bilateral symmetry. It hardly needs measurement. If something is askew we notice it, like the ears, for instance. Nevertheless a check was required, as what we take for granted or expect, we may not really understand. Merely asserting that bilaterality is so does not explain it in the least. All I wish to do here is to set bilaterality in the context of a comprehension of space, a wholeness, with its inevitable and inherent infinite elements. The form, the very structure, implies a particular specialisation of the placement of geometrical elements locally and at infinite distance. These elements are not merely theoretical add-ons. Accompanying the visible is always the invisible, but in our thinking we can begin to see it.

*Figure 4.37 Koala*
*Figure 4.38 A koala, (black & white)*
*superimposition of Figure 4.34*

*Figure 4.39 A wallaroo, superimposition of Figure 2.17 (p.21)*

Human beings too are unequivocally bilaterally symmetric, but while there is patent bilateral symmetry there is something else too. Is this a consequence of maturation, the intrusion of the personality into the basic architecture? Figure 4.40 is a photograph of the author. In Figure 4.41 the image is split, flipped and two lefts stuck together and two rights. Which of those in Figure 4.41 is the person in question? Neither, of course. They are distinctly different, and it is not just the haircut. Something gives us our specialness – an aspect of which is, I believe, revealed in the asymmetry of our physiognomy – and which we do not see to the same extent in the animalic world. Their specialness must come from a different source.

Is our specialness, our personality, embedded and revealed in this characteristic asymmetrical expression, which we glimpse when we depart from a strict geometric symmetry?

*Figure 4.40 The author, with a central line of symmetry*
*Figure 4.41 The author with two left halves and two right halves*

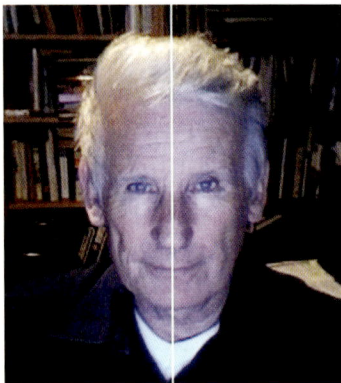

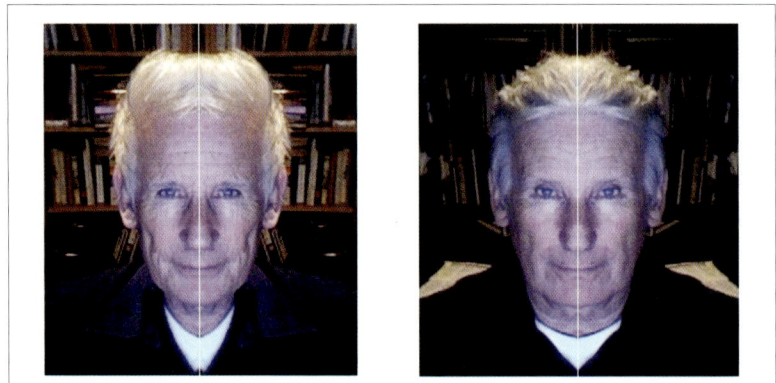

## 4.4 Rotational symmetry and forms of nature

This symmetry is represented in every flower form that has equispaced petals or sepals around a centre. The passion fruit flower has a twice fivefold symmetry. This symmetry is about a point but it is also within the compass, within the containment, as it were, of a line, and this time the line is the line at infinity (Figure 4.44). Again I use a dashed circle to represent the line at infinity, *o*.

To construct a basic figure out of the Desargues construction within which to understand rotation, begin with Figure 4.44, then place three lines, *x*, *y* and *z*, in *O*, and the three points, *X*, *Y*, and *Z*, on line *o*, the line at infinity (Figure 4.45). Select any triangle, *ABC* in the three initial lines *x*, *y* and *z*, (Figure 4.46). Now, to where will it transform? The vertices of the triangle go to *A′*, *B′* and *C′* still on *x*, *y* and *z* but now these have rotated according to where the points *X*, *Y* and *Z* have moved to on the line at infinity. In this case the triangles have moved clockwise. (Of course, there are two possible directions.) What we see then is the element (here a triangle) *rotating* around the centre point, *O*. The steps can be any equal angles.

What follows is, at one level, obvious to us all. For instance, if the sixfold symmetry of the lilies were suddenly to become eightfold, asymmetric or haphazard, we would surely notice! Much of nature's formal beauty and architecture we take entirely for granted. We expect symmetries such as those the lily and the rose give us. Fivefold is what apples simply do. No amount of Linnaean classification will tell us *why* the Rosaceae have fivefold symmetry any more than why sea urchins also do. Nor will what is presented in this book. The attempt here is to see if it is possible to put these structures into some kind of macro context, an overall structural architecture, rather than believing that the micro viewpoint solves everything.

*Figure 4.42 Passion fruit flower showing twice fivefold symmetry*
*Figure 4.43 Apple core*

*Figure 4.44 Rotation basics*
*Figure 4.45 Adding three lines and three points*
*Figure 4.46 Adding a triangle*

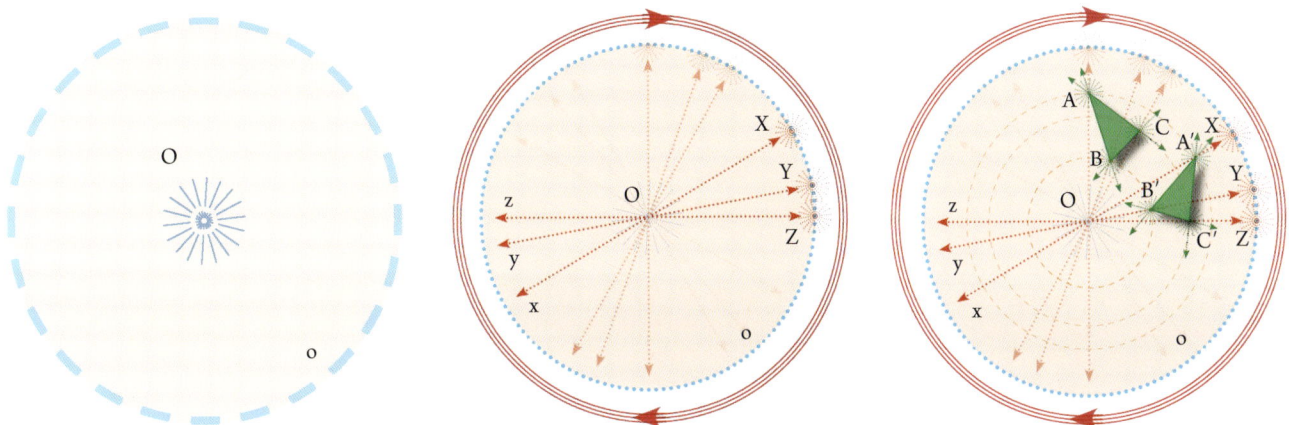

*Figure 4.47 Tea tree flower*

*Figure 4.48 General rotation construct*

## 4.5 Rotation in flower forms

There are a huge number of flower forms that exhibit a rotational format. It almost seems unnecessary to try to substantiate this rotational symmetry in nature, as it is so obvious. Or rather we think it is. But flowers don't rotate. What we actually see is some kind of *resultant* (the combination of two or more effects). Perhaps the petal layout is a resultant of an intermittently expressed rotation, a periodic appearance, a rhythmic breathing between manifestation and withdrawal, and even, if we take the notion of two points moving in opposite directions on the line at infinity, some kind of a fivefold standing wave around the centre, so that all we see is a phase appearance of an actual rotation. I suspect, in fact, that we are dealing with a double rotation; that is to say, that along the infinite periphery there is a rotation in two opposite directions, and what we see physically is the outcome of two rotations momentarily coming into phase – so to speak – and manifesting briefly.

Nevertheless, some examples need to be examined as confirmation of this apparent rotation, especially to find if there is an asymmetrical rotation lurking somewhere. An asymmetry would be revealed by any slight non-circularity of the flower form. This might be very hard to detect, and I am going to leave that for future work. The task is made difficult as by the time the flower is fully out and blooming, all sorts of influences from the environs have been at work on it, from caterpillars to storms, car fumes and chemicals.

There is one further point to consider. Before getting to the regular flower forms it is worth looking at a conceptual step towards this, in Figure 4.48. The general case is where each of the petals does *not* have a central axis, but is in itself asymmetrical. The special case is where the petals show bilateral symmetry and the whole blossom shows radial symmetry. There are a vast number of flower forms that exhibit this.

There are many flowers that have the kind of symmetry seen in spinning jasmine (Figure 4.49), probably a *Trachelsperman* or

*Jasmine.* To check whether these petals are truly rotating about a centre, we can map a series of points on each of the petals and see, after rotation about an estimated centre, to what extent they are thesame as each other as we go round.

First select a centre on the photograph of the chosen flower form, an estimate at best in this case, Figure 4.50. Then put in some reference circles about this centre at points that exhibit some distinction. Select one typical petal and draw in a single triangle on these circles. Since there are five equispaced petals, the the rotation angle is 72° (that is, 360/5). Plot five such triangles 72° apart. Rotate and adjust the whole complex so that the five triangles approximately cover the actual petals. We can see from Figure 4.51 that for the spinning jasmine the correspondence is not exact, but sufficient to suggest a correspondence.

Another example is, I believe, some sort of *Vinca* or periwinkle, where this seeming rotation is apparent. Again, select an estimated centre of the flower (Figure 4.53). Put in some reference points on a typical petal. Let circles, centred on the estimated centre, pass through these points (three in this case, one of which is on the small central fivefold polygon, Figure 4.54). Now let the imposed complex rotate around in multiples of 72° around the centre (Figure 4.55). This shows the discrepancy between the actual spacing and an exact spacing of the petals.

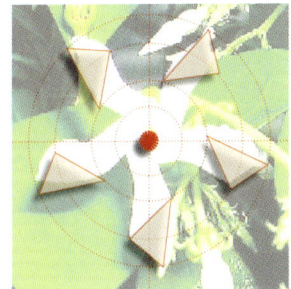

*Figure 4.49 Spinning jasmine*
*Figure 4.50 Centring*
*Figure 4.51 Adding five elements*

What is the nature of the 'error' or difference between actual and ideal spacing? I selected a typical point on each petal and measured the angle from a base line going anticlockwise. If we simply sum the angles between petals we have to get 360, which tells us nothing except the average. That is, 59 + 76 + 80 + 74 + 71 = 360. But if we square these values, sum them, take the average and then take the square root, then there is a weighting given to each measured number and an average obtained which reflects the variance from ideal:

$$59^2 + 76^2 + 80^2 + 74^2 + 71^2 \;=\; 26174$$
$$26174 / 5 \;=\; 5234$$

So the new average is $\sqrt{5234} = 72.3$. This is reasonably close to, but not exactly, 72°. (To show that this is significantly close, we do a little test with wildly different figures, say, 10, 10, 10, 10 and 320, which also sum to 360, and still average at 72. Squaring these values gives 102800 and then 102800 / 5 = 20560, and $\sqrt{20560}$ = 143.4, which is nothing like 72.)

If rotation is genuinely true here, then I have established to my satisfaction that the overall construct needed is of the dual nature described in Figure 4.44; there is the need for a centre, *O*, in the very middle of the organism, and a periphery, *o*, in the infinite. The conceptual construction is inherently incomplete without both these entities, the central and the infinite. To put it another way – the plant, the periwinkle, is incomplete without the cosmos.

## 4.6 Rotation and bilateral symmetry combined

There is a further step which belongs here. Can we have *combined* symmetries? For rotation symmetry can be, and often is, combined with bilateral symmetry. Consider these two symmetries together as in the amazing passion fruit flower. In fact this appears to be the more prevalent instance. Another example, this time with five petals also shows bilaterality within each petal (Figure 4.57).

*Figure 4.52 Periwinkle*
*Figure 4.53 Periwinkle centring*
*Figure 4.54 Periwinkle with triangle*
*Figure 4.55 Periwinkle matching triangles*

Subject as they are to so many external vagaries, we can't expect the petals to be exactly geometric when they are photographed. A further example is a kind of daisy. In this case I have merely overlaid an arbitrary element thirteen times for each of the single petals, and compared (Figure 4.59). We easily see how much the real thing has strayed from some ideal, yet there is still a thirteenfoldness. To conclude, here is the outside of a passion fruit showing a well defined sixfold rotational symmetry (Figure 4.60).

Is there a double rotation in these examples? Is there a sort of balance of activity (as it were) in each direction on the infinite line? Is the rotation symmetry or the bilateral symmetry the primary gesture? Our explorations seem to throw up more questions than answers.

*Figure 4.56 Passion fruit flower*
*Figure 4.57 A five-petalled flower*
*Figure 4.58 Daisy*
*Figure 4.59 Daisy with regular thirteenfold overlay*

## 4.7 Translational symmetry in nature

We might be of the view that the simple repetition of an element (perhaps a triangle again) in a linear path is easy and straightforward to understand. To me it is the least apparent compared to the more or less straightforward structures that demonstrate bilateral and rotational symmetry.

Translational symmetry is the basic symmetry of the mineral world. The other symmetries are also a part of mineral architecture but it seems to me that the primary gesture is one of congruency and repetition. Is one atom not the same as its neighbour if of the same element? Here we have the crystalline, the rigid, the repetitious, linearity, fixity and rectangularity. This stark, precise regularity is revealed in our modern culture. Look at almost any cityscape and we see skylines reflecting this obsession with the stark verticality of a vertically-oriented crystal garden. Try to find a curve that is not part of nature! Do we, then, have a kind of mineral consciousness today? It would seem that on an architectonic scale we certainly do. The mineral does indeed 'do' rectangularity. But it is far from limited to this – most crystal forms are not riddled with right angles, but still have a definite regularity.

*Figure 4.60 Passion fruit*
*Figure 4.61 Linearity on a high-rise block*

*Figure 4.62 Cityscape*

*Figure 4.63 A crystal-scape*

*Figure 4.64 Calcite*
*Figure 4.65 Garnet*
*Figure 4.66 Quartz*
*Figure 4.67 Pyrites*

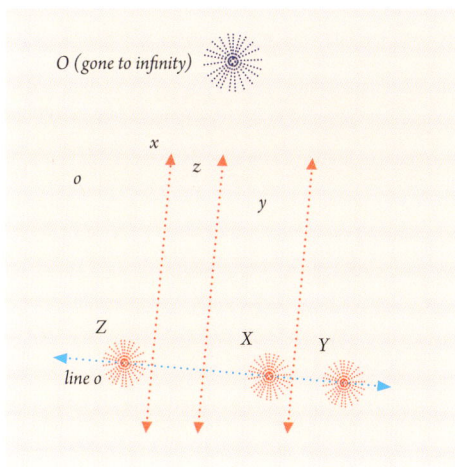

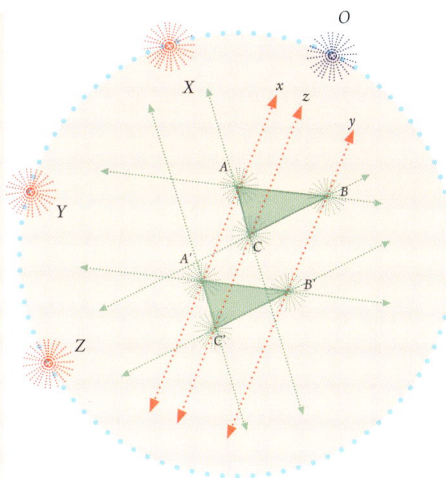

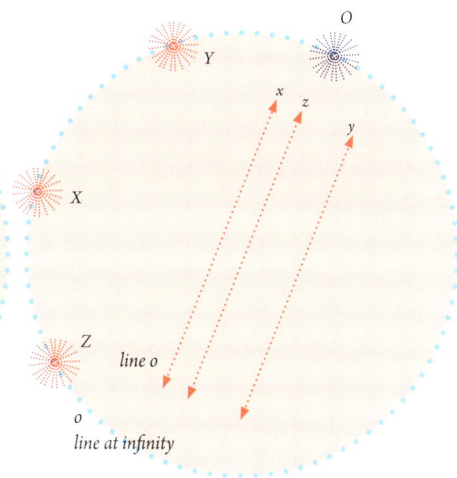

We see this regularity for instance in calcite, garnet, quartz, tourmaline or variants of pyrites.

Do the surface lines on many crystals indicate the repetitive layering, and hence the translation of a particular form – rectangular for pyrites, triangular for tourmaline, hexagonal for quartz? In this mineral reality it is not all perpendiculars.

Bilateral symmetry required a local line, *o*, and a point at infinity, *O*, as well as three lines *x*, *y* and *z*, meeting in *O* at infinity (which makes them parallel), and three points *X*, *Y* and *Z* on the local line, *o* (Figure 4.68). Translation symmetry requires something further. The line, *o*, must be moved to infinity as well (there is not much left on the page). I have again taken the liberty of using a dotted circle for this line at infinity (Figure 4.69).

*Figure 4.68 Bilateral symmetry*
*Figure 4.69 Translation symmetry*
*Figure 4.70 Translating triangles*

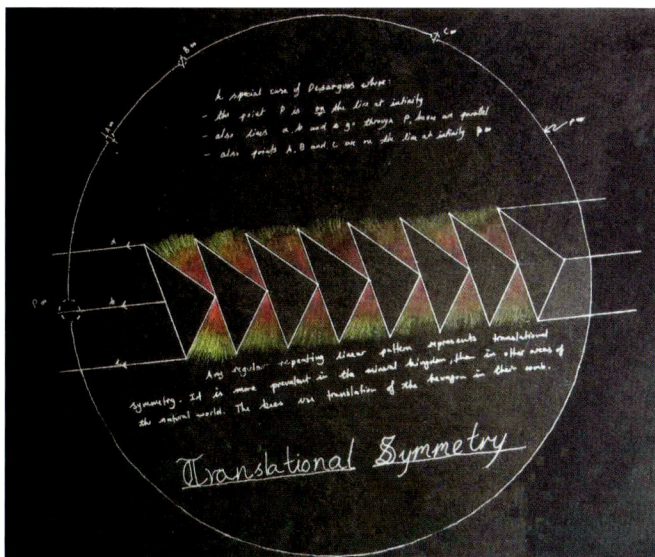

*Figure 4.71 Translation [Marco]*

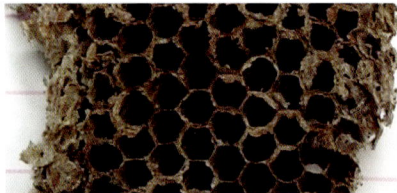

*Figure 4.72 Windows at a Catholic College (in Melbourne, Australia) designed by Walter Burley Griffin.*
*Figure 4.73 Australian native wasp paper cells*
*Figure 4.74 Tessellation (or translation) of hexagonal scales on the body of a fish*

Now triangles *ABC* and *A'B'C'* are still on lines *x*, *y* and *z*, but merely slide along them, like on railway tracks (Figure 4.70). *A* goes downwards to *A'* along line *x* (say) effectively rotating about *X* on *o*. This is because parallel lines can be thought of as rotating about their common point at infinity. These constructions are easy to draw even if they are not that easy to grasp.

We can see examples of translation in architecture, for example in the windows at a Catholic College in Melbourne. We can also see examples from the insect world: bees make hexagonal cells, and wasps make paper cells that are about half the size of those of honeybees. There is a tendency toward a tessellation on the body of the little fish found on Zenith Beach in New South Wales. The mostly hexagonal scales tend to repeat across the skin.

Countless human artefacts manifest this translation symmetry. It will not occur to most of us that inherent in these structures is not only the centre of every repetitive element but also a single infinite line or a periphery.

## 4.8 Centre and periphery and two kinds of measure

If we really are going to apply geometry to nature then we should apply it as fully as we possibly can, and not just adorn nature with a few incidental bits and pieces of this ideal mathematical world for the convenience of description. If we take geometry seriously we may evoke new viewpoints towards the physical world of our senses.

Looking at a particular object, it may seem obvious where the centre lies, as for instance in the sea urchin shown in Figure 4.75. Is the centre really there? At best it is an estimate for an abstraction which we feel may be helpful. And to estimate the *centre* we must consider the *periphery* (geometrically in the infinite, as urchins suggest a rotational circular symmetry). We are usually not consciously aware of the periphery, but are convinced that we can actually see the centre. It is inferred at best. So we should perhaps realise that radiating from any central feature are numerous rays. In the urchin these are visible as the fivefold radial symmetry so characteristic of this species.

We must allow that for every centre, there is a related periphery. The infinite line, where that is the periphery, is all around us. But if we regard it simply as a fiction, then so is a centre, since *both* are geometric entities. The centre is simply an incomplete idea without the periphery. The inner geometrician in us knows that unless this line is considered, we have merely a half-truth.

If the urchin is truly fivefold, then the angle between the segments would be 72° and between the edge and the line through the middle of each segment will be 36°. That is, there are equal *angles.* This is *angular* measure. The conventional units, the degrees, do not matter at this point.

What does this say about the line at infinity? What occurs along it? Along this line there are equal *distances.* There is an equality of *distance* measure. For equal angles must mean we traverse equal distance along this line – however far away it is. (This will be discussed more in Chapter 7 in relation to the three kinds of linear measure.)

So we have two qualitatively *different* kinds of measure, and we arrive at *both* kinds of measure in the same object. This is reflected in the elementary geometrical instruments of protractor (for measuring angles) and ruler (for distance) with the everyday units of degrees for angles, and centimetres and millimetres (or inches and tenths or sixteenths), which are simply conventions used for convenience.

*Figure 4.75 The centre of an urchin*
*Figure 4.76 Centre and radiants*

## 4.9 Two kinds of two dimensions

The two extreme geometrical elements are the plane and the point. Almost all the geometry shown so far has been *in the plane* – be it a sheet of paper, blackboard, book or monitor screen. We commonly think of this as 'two dimensional', and it most certainly is, of course. But if we treat the elements of geometry seriously and equally, is this all there is? What about the point? We can have a whole geometry *in a point!*

There are equivalents in the point-wise to all the symmetries so far examined in the planar. There must be, as the geometry is consistent. We must simply take the notion of *duality* or *polarity* seriously (see the end of Chapter 1 for more detail on this). However, I will not attempt to explore all these, but will simply show a few examples from nature. To begin, here is a graphic showing both cone and circle together. The circle in the plane π is built of tangent points and lines; it has its counterpart in the full cone in the point *P,* built of both planes and lines (Figure 4.77). This picture is fully polar (I use *polar* for three dimensions and *dual* for two dimensions). Whatever one can say about constructions in a plane, there will be equivalents in the point.

Taking the germ *point* in all earnestness, we see organisms evolve from it, growing with incredible regularity into the spaces of our world. The seeds of the lawn I planted tend to the point-like. Our own physical body was dependent on the magnificent mystery of the cell growing into embryo.

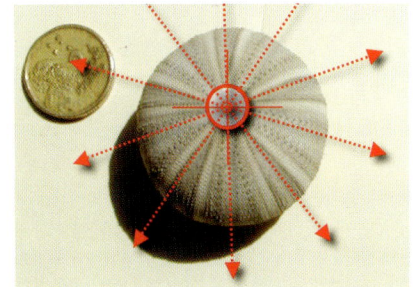

*Figure 4.77 Cone and circle of lines*

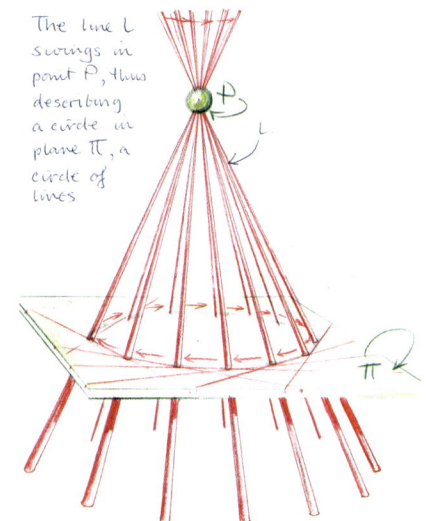

The line L swings in point P, thus describing a circle in plane π, a circle of lines

*Figure 4.78  Pearly top shell, Tectus*
*Figure 4.79 Two shells showing different cone angles*

Where does nature emphasise this point-wise aspect? There are radiations of all sorts. We only have to look at sea shells to see this. The cone angle is approximately 60° in the pearly top shell, *Tectus* (Figure 4.78). In other shells, the cone angle can vary from wide to quite narrow. Two more examples are illustrated in Figure 4.79. It is not difficult to mirror such forms in a drawing.

We can go further. If the way in which the spirals going around the cone are taken into account, and a full circle of shell growth is assumed for each step along the central axis then it is possible to sketch something of the growing form (Figure 4.81). The two dimensional structure becomes self-evident in this sketch as the planes and lines through the apex are clear. The spiral has been imposed on this – which actually makes it three-dimensional.

In the plant world we see many an instance of what appears

*Figure 4.80 Drawing of a spiral in a cone*
*Figure 4.81 Sketch of a shell form.*

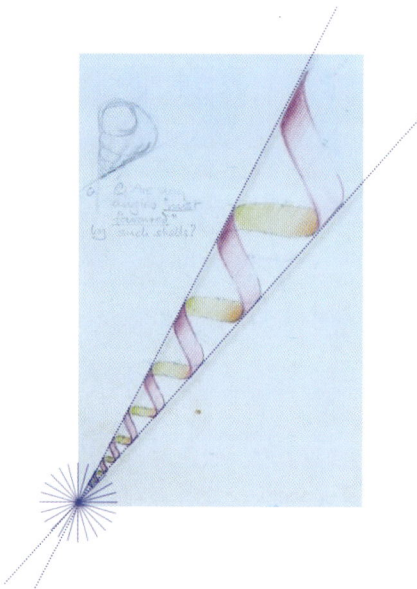

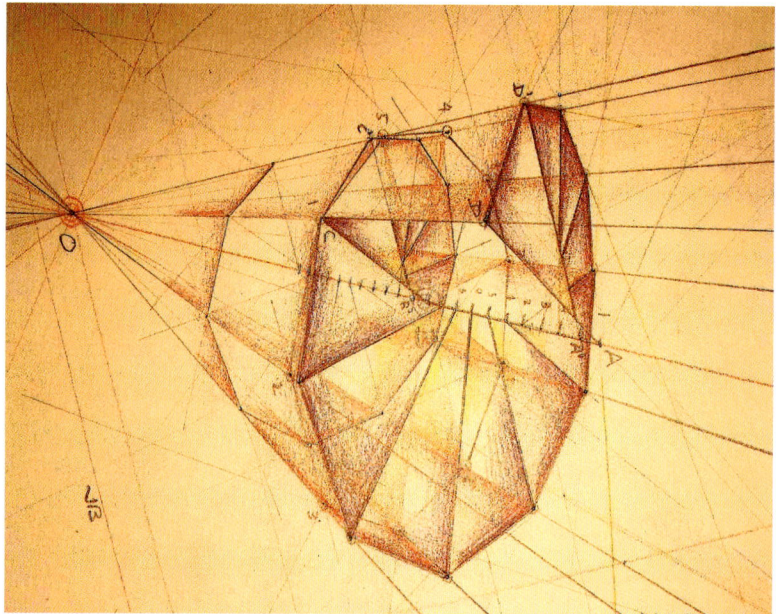

*Figure 4.82 A large-leaved plant of the Agavaceae family.*
*Figures 4.83 and 4.84 These flowers show a similar radiation from a point*

to be a radiation from a centre of sorts. We're all familiar with the dandelion seed head, but this effect occurs on a number of scales. Figure 4.82 shows a large-leaved plant of the Agavaceae family which appears as a large three-dimensional star. A similar morphology presents itself in the glorious red flower head, Figure 4.83, and this is almost spherical too.

Another little graphic exercise that is straightforward to do is the polar form of the pentagon and pentagram in both point and plane (Figure 4.85). This represents the dichotomy of form in point and in plane, and shows that, despite these two viewpoints being quite unlike, they are intimately related.

*Figure 4.85 The polar form of the pentagon and pentagram*

# 5. Asymmetrical Rotation

As we've seen already in Chapter 2, if the line in a basic geometrical structure is infinitely far away, we get translation and rotation (see sections 2.5 and 2.6). If the line is central, we have bilateral symmetry (section 2.4). But can we get situations where *both* central point and outer line are relatively local, thus resulting in an asymmetrical rotation?

I had not expected to find this, although I was curious as to whether it existed in nature. The first time I noticed an instance was with a leaf fan (as mentioned in section 2.6). I then began to see it more and more, for instance in the Botanic Gardens in Sydney. Figure 5.1 shows the plant I first saw, and Figure 5.2 shows a local sample. To what extent does this tenfold leaf system mirror an asymmetric geometry? I only did an approximate analysis – enough to satisfy myself that there was something significant going on here.

## 5.1 Asymmetrical leaves

In order to understand the geometric analysis, I will demonstrate this construction from scratch.

> First select any point and a line (not through the point) (Figure 5.3).
> Draw a horizontal line through point *P* (Figure 5.4).
> Now draw radiants from point *P*, every 20° (Figure 5.5). The ninth point will be at infinity.
> Insert a point *A* on one of the radiants and draw a line *a* through *A* and point *1* (Figure 5.6).
> Now draw the next point-line pair: point *B* is where line *a* crosses the next radiant; line *b* connects *B* with the next point *2*. Now the pair begin their rhythmic alternating 5.7). movement and from there continue all around (Figure

An elliptical curve emerges (Figure 5.8).

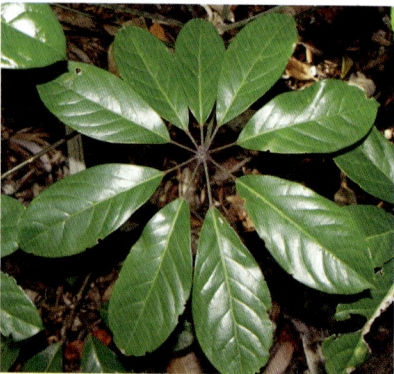

*Figures 5.1 and 5.2 Leaves showing an asymmetric rotation*

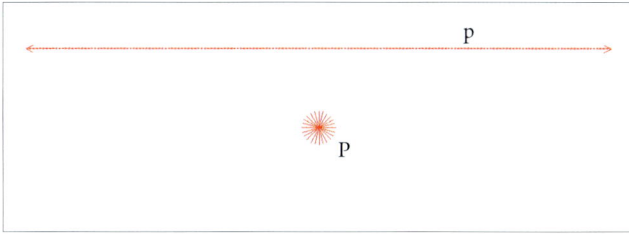

Figure 5.3

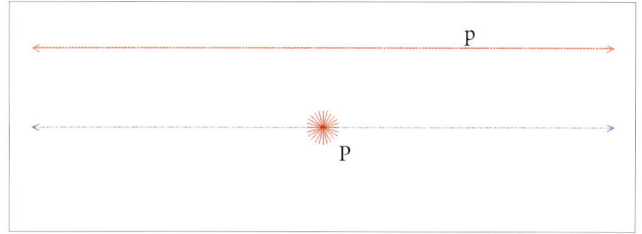

Figure 5.4

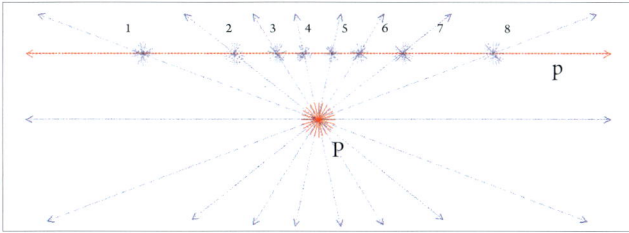

Figure 5.5

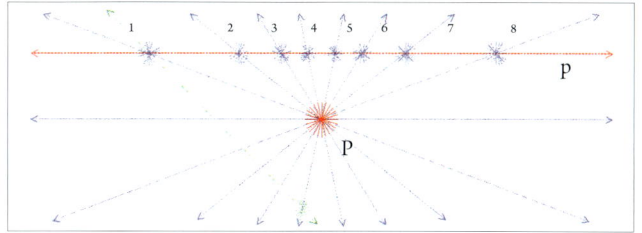

Figure 5.6

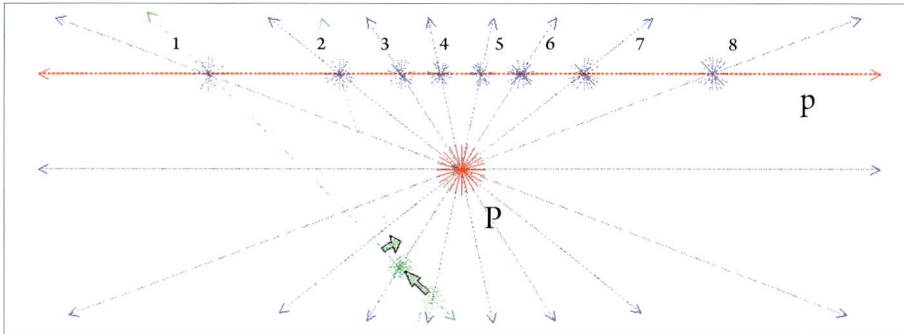

Figure 5.7

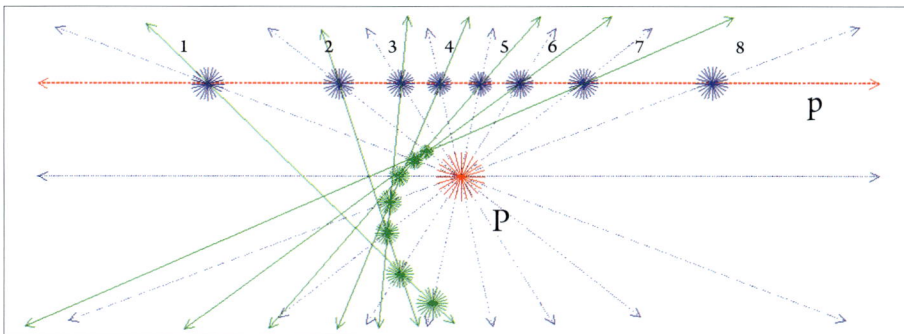

Figure 5.8

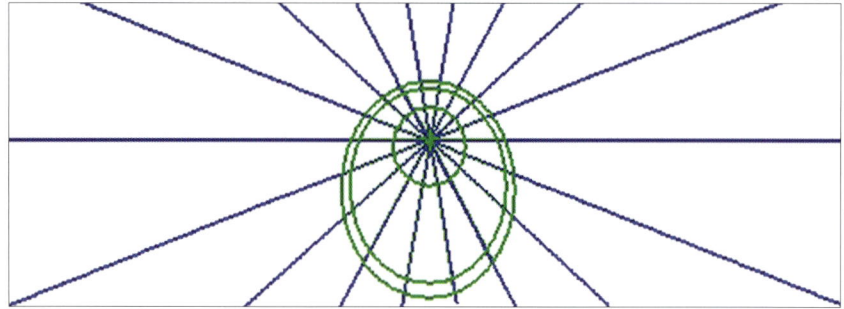

Figure 5.9

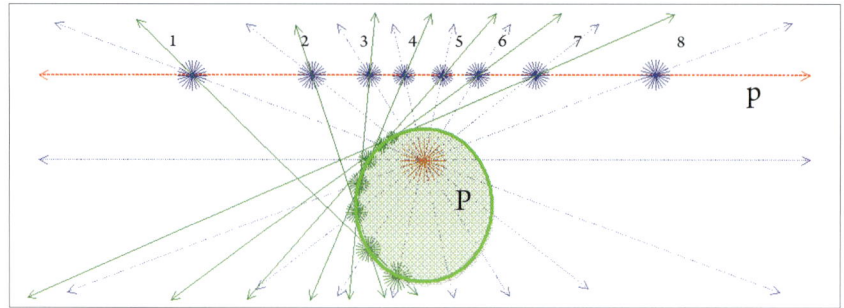

Figure 5.10

Figure 5.11

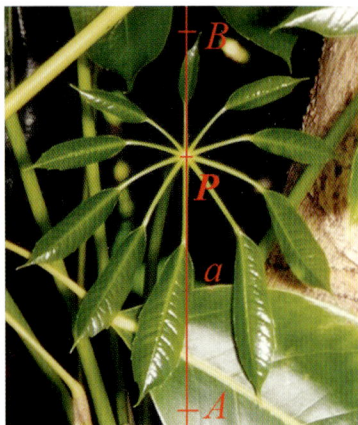

Figure 5.12

The complete curve can be drawn in (Figure 5.9).
Now some of the background construction is omitted
    (Figure 5.10).

Returning now to the leaf fan, assuming that the geometry *is* an asymmetric rotation, we need to make a similar construction but starting with the actual leaf. The leaf system I chose was tenfold.

Set a vertical along the estimated long axis of a selected
    tenfold leaf complex.
Estimate centre of radiation of complex, *P* (Figure 5.11).
Estimate a centre line, *a*, through the ends of the closest
    and furthest leaves, marking these end points *A* and *B*
    (Figure 5.12).
There is a centre point, *P*, two points on either side, *A*
    and *B*, *not* equidistant from *P*. To find the harmonic
    point to *P*, draw any two lines through point *A* and
    one through centre *P* crossing them. At these cross-
    ing points draw two lines to point *B*. This is called a
    harmonic quadrilateral. The second diagonal of the

Figure 5.13

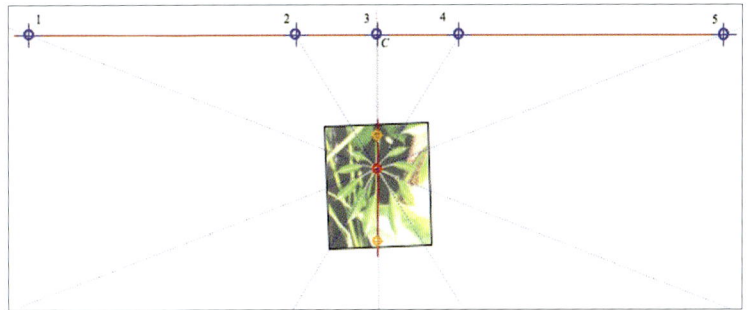

Figure 5.14

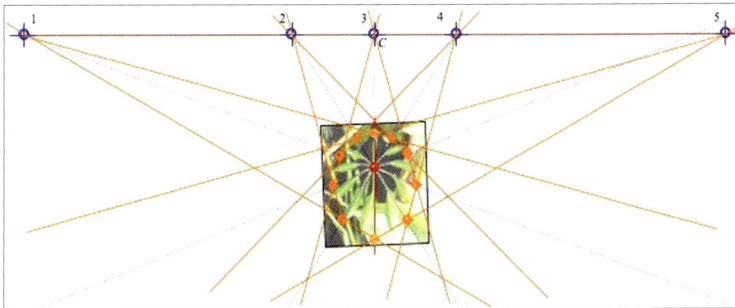

Figure 5.15

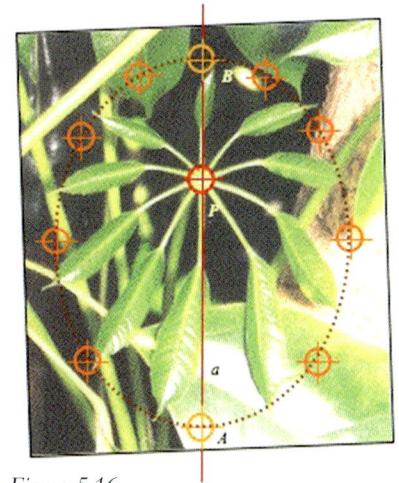

Figure 5.16

harmonic quadrilateral will lead directly to point C. This point C is harmonic to the centre P (Figure 5.13).

Point C is also the position of the peripheral line p that is needed, at right angles to line a. Note that the peripheral line is not in the infinite.

Having found an estimate of the core architecture we can now try to relate this to an asymmetrical rotation of the leaves. From the centre let the stems ray out, through the leaves, to the line p. There are ten such leaves. Assuming that they are equispaced about this centre, they would be 36° (360/10) apart (Figure 5.14).

Using this equiangular assumption it is possible to draw tangents from the points on line p (a circling measure) to surround the overall form that the leaves appear to take (Figure 5.15). Since it is assumed to be an asymmetric rotation, these tangents surround the overall form with an *ellipse*.

Figure 5.16 shows the supposed surrounding elliptical form. There is a reasonable match, I believe. This gives some validation

*Figure 5.17 Azalea*
*Figure 5.18 Azalea with ellipse*
*Figure 5.19 Harmonic construction*

to the notion that at least some plants respond to an asymmetric geometry.

Note that this construction has a *local* line as 'periphery' as well as having a local 'centre'.

## 5.2 An asymmetrical flower

Usually we think of flower forms as regular, that is, circular. But I noticed one flower that did not quite seem to be pure rotation. There could, of course, be many. But the azalea flowers in a friend's garden struck me as not quite circular, so I decided to check (Figure 5.17).

Was there a bilateral aspect in each petal, overall bilateral symmetry *and* rotation, all thrown in together? Why was I suspicious of this thing? I noticed that one petal – at the top – was slightly larger in width than was seemly. Another clue was that the colouration varied around the petals, with a concentration of deeper colour at the top and also a speckling not visible in the other four petals.

For pure rotation a circle would surround the petal tips – here then there should presumably be an *ellipse*, albeit of slight eccentricity. I established such an *estimated* ellipse in the background image simply because the scale gives a better chance to approximate this form around the petal periphery.

Estimate a centre point, A', and then estimate a vertical axis.

Estimate an ellipse form over the petal tips (I used an ellipse tool on the computer). This gives an approximate top, O, and bottom X'. Both points must lie on the vertical axis selected (Figure 5.18).

Now find the *harmonic* point, A, to the centre, A', with respect to the top and bottom points, O and X. This is best done on a large sheet of paper (Figure 5.19). The intersection point, A, in this case is well off the page. Hence we are reduced to calculation through the *cross ratio* process that is inherent in the harmonic quadrilateral construct. By calculation, the point A is 1233.7 mm from the middle point A' if XA' = 83.7 mm, A'O = 73.7 mm and the cross ratio R = −1. That is, 1160 + 73.7 = 1233.7 (see panel opposite for calculation).

So the outside line governing the overall form needs to be through point A at right angles to the (almost) vertical axis.

The next step is to check the petal *angles*. If there was pure rotation then the angle between each petal should be

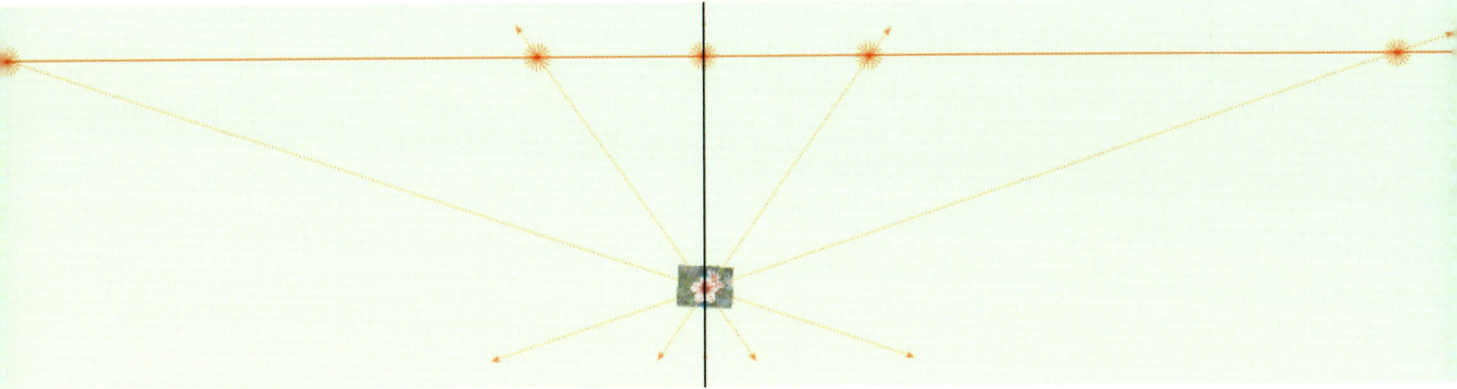

Figure 5.20 Finding the points on the circling measure

72° (360/5). This implies a *circling measure* or asymmetric rotation symmetry. For this the construction has to be even larger, for the polar line, $a'$, through the point $A'$ has to be long enough to contain the five points in circling measure. If this is constructed we would get something like Figure 5.20.

This is the overall Azalea flower structure from a planar perspective, showing the geometric elements determining the five petals profile.

**Finding a harmonic point using cross ratio**

Let the cross ratio $R = -1$ and $x = OA$

Then scaling from the image:

$$XA' = 83.7 \text{ mm}$$
$$A'O = 73.7 \text{ mm}$$

Then $XA = 83.7 + 73.7 + x$

One form of cross ratio where $R = -1$ is

$$R = (XA / OA) / (XA' / A'O)$$
$$-1 = (XA / OA) / (XA' / A'O)$$
$$-1 = ((83.7 + 73.7 + x) / x) \times (73.7 / 83.7)$$
$$-1 = ((157.4 + x) / x) \times 0.8805$$
$$-x = (157.4 + x) \times 0.8805$$
$$1.1357x = 157.4 + x$$
$$0.1357x = 157.4$$
$$x = 157.4 / 0.1357$$
$$x = 1160$$

As $x$ is OA the distance is 1160 mm

## 5.3 *The great universe*

It is apparent that the mathematical, geometric framework in which this flower exists is rather large. I contend that this architecture is as much a part of the plant as is the material petal. If the maths really applies to physical phenomena, as is simply taken for granted in much of physics, why not here too? One could even ask if some kind of field structure, a morphological field, pervades the entire space in which our little flower exists?

Are we justified in imagining that the plant form, from the *formative* point of view, inhabits a *far* greater space than is usually imagined? I believe so. Perhaps such fields can even be accurately configured. I found support for this view in Steiner where it is stated 'All that takes place in the plant is an effect of the great universe' (*Karmic Relationships*, Vol. 1, p.14). From the geometric formative point of view at least, this seems to me to be valid.

# 6. The Orientation of the Line

Do any of the geometric elements have an orientation, and if so, how? Are they independent of such dictates, and is any position as valid as another? Obviously a point cannot have a particular orientation. But lines could point somewhere: they have an orientation.

## 6.1 The mineral realm

With mineral crystals there seems to be no orientations at all. In themselves they are often highly structured, but within a mass of crystals their orientation appears to be random. There are some exceptions: some formations like the basalt pillars of the Giant's Causeway in Northern Ireland, Fingal's Cave on Staffa in Scotland, or Tasman Island off Tasmania, show a verticality, they could be said to relate to the centre of the earth.

However, as an aside not strictly concerned with orientation, we can consider the regularity of crystal structures in themselves to be forms between the inmost point and the outermost periphery. Any line (or axis) is between the crystal's centre and the infinite sphere. As this infinite sphere completely surrounds us, the lines can lie in any direction. Parallel lines meet in this plane at infinity. For instance, the three sets of parallel lines of a cube all meet in the *same* plane, the plane at infinity. Hence the *centre* point of the cube (or any crystal) has a counterpart in the *plane* at infinity.

Orientation is not an issue here. Note how a *family* of rectangular prisms (for instance, pyrites), while clearly having geometric centres that are totally distinct, all have the *same* plane at infinity as far as their geometry is concerned. The edges of each separate crystal block, a rectangular prism, point in *different* directions, yet the vanishing points of all these edges end up in the *same* place, the infinite or absolute plane.

*Figure 6.1 Crystals showing a random orientation*
*Figure 6.2 Pyrites and its parallel lines that meet in the plane at infinity*

*Figure 6.3 Layers of cloud*
*Figure 6.4 The horizon looking east over Sydney to the Tasman Sea*
*Figure 6.5 The Blue Mountains in New South Wales: the relatively flat plateau has deep, rugged valleys*

However, in the inanimate mineral world there is often a layering, a horizontal plane. It can be seen in the layering of clouds in the atmosphere, in the flat surface of calm water, and of course in the horizon itself. And we often see the horizontal layering in the strata of rock formations laid down millennia ago. In many a road or rail cutting the laminar presentation of the rocks is patent, as well as a few undulations, perhaps when the materials were more plastic.

## 6.2 The plant world

In the abstract generalisation of 'space' nothing singles out any particular direction, but when we look at the world we know, this is pointedly untrue. Consider the sun and the earth and the line between their centres. (This is the line of attraction between two centres of gravity and is presumed in many astronomical calculations.) The plant world takes this line very seriously. Part of a plant grows towards the light while another part grows towards the centre of the earth. If one were to characterise their general growth, they grow straight up and straight down.

This orientation appears to be fundamental to what Goethe called the *archetypal plant*. There is something awesome in an old forest. (Is one among great beings in such a forest? Is Tolkien's world of ents so far fetched?)

If we imagine all the plants around the world, we would have a vast number of trunks, stems and straight lines aligned between earth's centre and periphery, generally at right angles (or perpendicular) to the surface of the earth. This vertical orientation is important to the plant and is part of its very nature.

The plant is a creation of the earth *and* the light: one aspect drives into the earth, tending towards the earth's centre, the other strives towards the light represented by the sun. We can imagine the plant world around the earth akin to the hairs on our head. The hard surface of the earth is a bit like a skull. Even on a small scale, the vertical tendency dominates, as can be seen by shoots growing straight up out of their 'earth', the horizontal bough from which they grow (Figure 6.7). This living base does

*Figure 6.6 These trees show the perpendicular tendency of the plant world*
*Figure 6.7 A stem growing vertically out of a local banksia tree*
*Figure 6.8 Grass tree spikes with their seed heads on their spear-like stems, against the backdrop of garden railings illustrate the vertical*

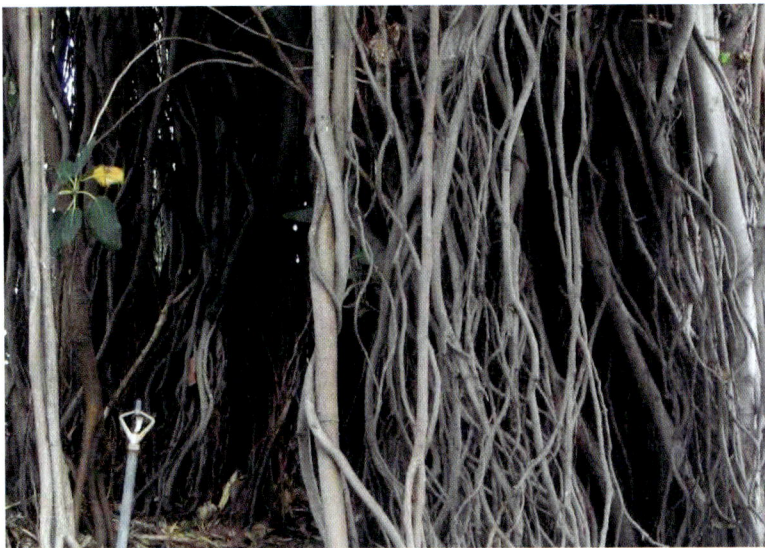

*Figure 6.9 Aerial roots of fig trees in New South Wales*

*Figure 6.10 This damaged tree has literally taken a U-turn towards the light above, showing a serious striving towards the light*

*Figure 6.11 Mangrove roots in the tidal reaches of a cove in Sydney, Australia*

not hinder the stem in rapidly finding the vertical, and even its own flower head can be nice and vertical, parallel to the woody stem itself.

Another aspect is shown by many of the huge fig trees on the East Coast of Australia. These have aerial roots, which grow down from branches fairly high in the tree, forming what seem to be a host of supporting columns.

Some mangroves have numerous roots that grow both upwards and downwards. Upward growth is necessary as the roots need to be exposed to the air as the swampy conditions prevent the roots breathing.

While the predominant orientation of the plant world is vertical, we can find a horizontal orientation in some plants. For instance, there is an Australian weed where the foliage comes away from

*Figure 6.12 Asparagus plumosis, a weed, showing a secondary horizontal tendency*

the stem in virtually a horizontal plane, or the great cedars show something of this in their foliage too. But this horizontality appears to play second fiddle to that of the primary verticality of the plant.

Seen from *above,* the plant world generally demonstrates a rotational symmetry about the central vertical axes. We see a point or centre about which the whole plant seems to revolve.

*Figure 6.13 Pines in Rishikesh, India. The branches branch out from the dominant vertical [Madelaine]*

*Figure 6.14 Fungi often grow in layers on dead and dying wood. These are in the Old She Oak Reserve in Sydney*

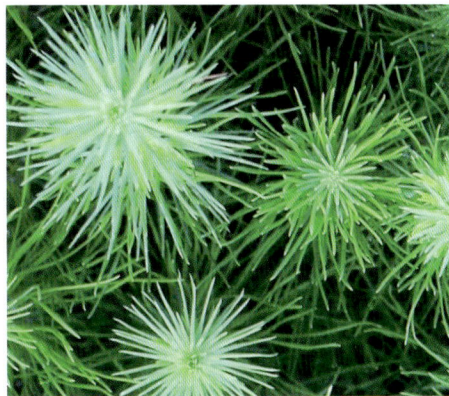

*Figure 6.15 The Monkey puzzle tree has branches that spread into a series of parallel planes as they reach the limits of their canopies*
*Figure 6.16 Look down from above on young alders,* Alnus glutinosa, *in a nursery*

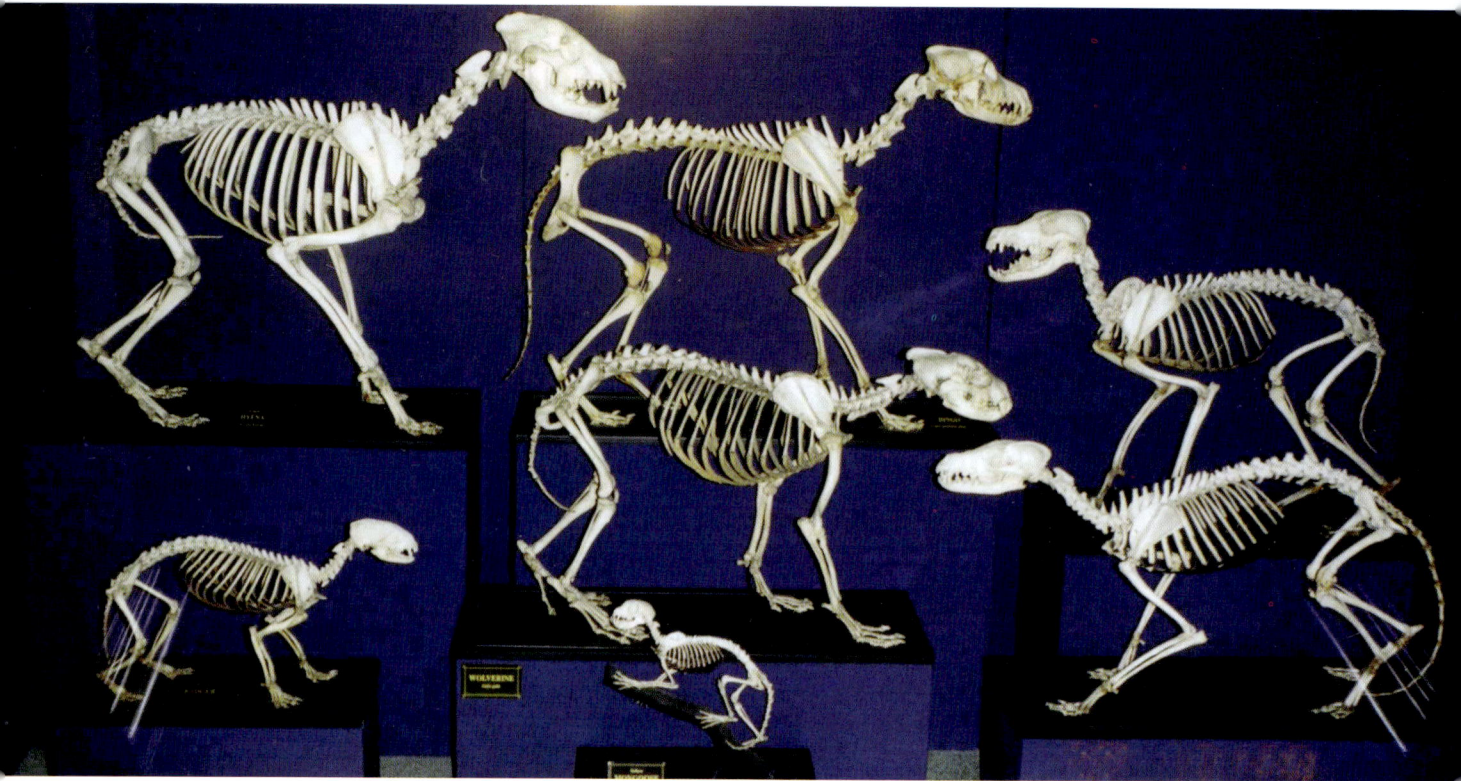

## 6.3 The animal world

When we come to the animal world, we find the predominant orientation is horizontal. While one could argue that it is no wonder that a fish is horizontal, for it floats; nevertheless the horizontal appears to hold sway in the majority of the animal world. Most mammals meander or race around on four paws, and their movement is largely in the horizontal plane. I suspect the stance has as much to do with the potential for movement and sentience, in that if a creature can move it needs some sense of where it is going!

*Figure 6.17 A number of mammal skeletons in the Australian Museum, Sydney*
*Figure 6.18 Echidna, and Australian spiny ant-eater*
*Figure 6.19 A rhinoceros*

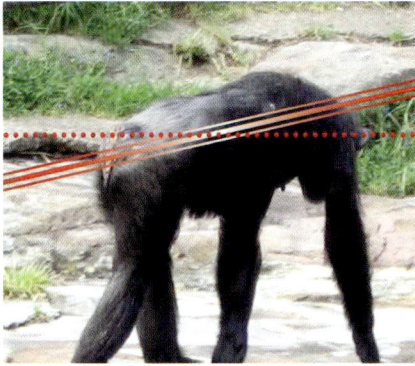

So, to a large extent the entire animal kingdom has a dominant horizontality. Seen from the side, the spine wavers above and below a horizontal line, the rump lower, the back arching, the shoulders lower, the head raised, the muzzle lowered, or sometimes the reverse of this. There is almost a rhythm in the undulation. From the top, however, the view is of the straight spine.

A nice little local example from New South Wales is our docile egg-laying spiny ant-eater, the echidna. The top view shows an obvious longitudinal straightness and a symmetric balance on either side of this line (a mirror reflection). Even though it is merely a single line, I do not think this simple picture of 'geometry in action' should go unremarked, for it concerns a whole kingdom of nature and is an essential part of its character and basic plan.

Even the spinal angle to the horizontal of the chimpanzee is not great in normal locomotion (Figure 6.21) as it knuckles along at our zoo. It is as if the horizontal stance has taken a small movement towards the vertical (about 10°).

*Figure 6.20 Antelopes*
*Figure 6.21 The chimpanzee does not show much deviation from the horizontal in normal locomotion*
*Figure 6.22 Even the kangaroo, despite its stance, is closer to the horizontal than the vertical*

## 6.4 The upright hosts of the earth

The human being, however, shows a tendency to verticality, which sets the human world apart from the animal. Straightness and uprightness is the fundamental gesture of the human being. Viewed from the side, the human spine wavers back and forth about the vertical, like the animal's wavers about the horizontal. However, viewed from the front (or behind) the spine is a straight line in the human body.

The vertical characteristic of the human being is like that of the plant kingdom. But is the one perhaps an *inversion* of the other? One hint of this is that our reproductive organs are 'downstairs', below the belt, while those of the plant are seen as 'upstairs'.

It is interesting that artists have seen this verticality with some clarity and expressed it over and again, from the two-thousand-year-old figurine of Etruscan times to more recent sculptures by Giacometti. Perhaps this verticality expresses something of an emerging individuality.

*Below, from left*
*Figure 6.23 A copy of an Etruscan figurine. The line in this figure is barely interrupted except for two emphases that do not detract from the primal gesture of uprightness*
*Figure 6.24 Alberto Giacommeti, Grande Femme III*

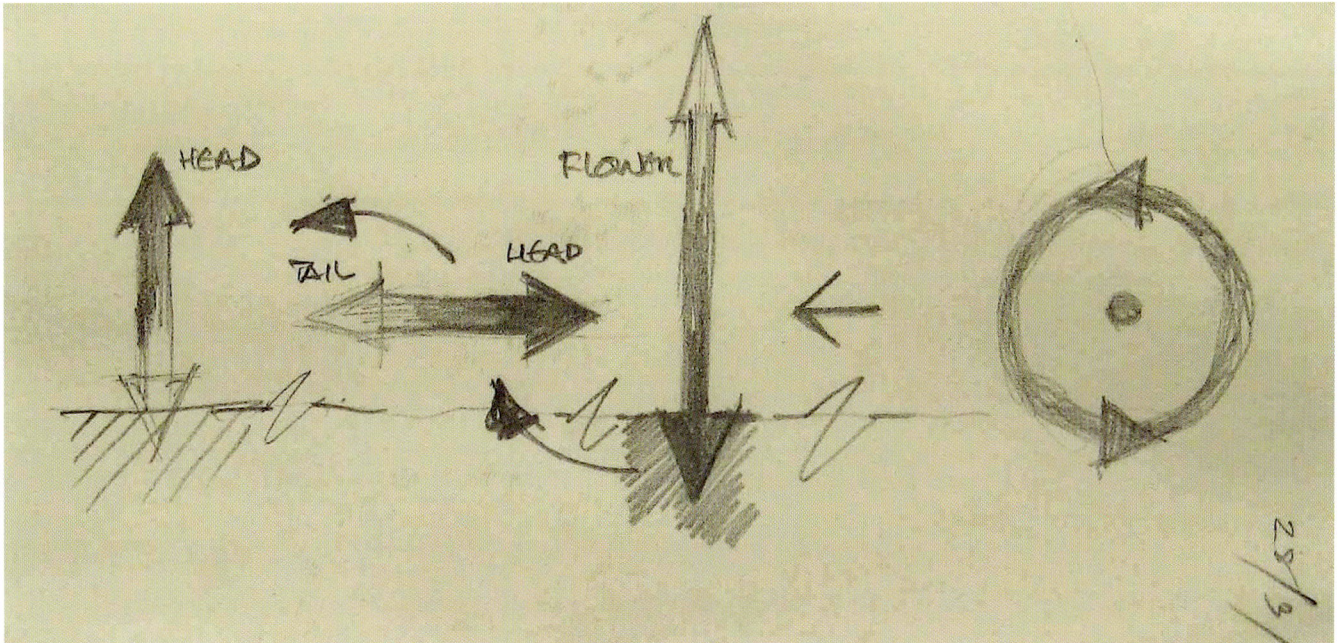

Figure 6.25 The sequence of orientation (right to left) from mineral to human

## 6.5 Summary

In looking at the different orientations of the kingdoms, it may seem that we have been discussing the obvious. Despite the fact that we take it for granted, I do believe there is some significance in the obvious.

The human being stands upright, directed both upwards and downwards. We look up to the sun, have our feet on the earth. Our inner being strives towards the light, our earthy being has its feet on the ground.

The animal has a horizontality, moving forward. Its world is basically parallel to the earth's surface. This is the world of species, each with a specialisation adapted to its environs. (This is quite different to the human being who has no such specialisation and is not bound to one particular environment.)

The plant has verticality, an uprightness of an entirely different nature to the human. One could say it is directed downwards with the root aspect and upwards with its fruiting aspect.

The mineral world is something of a mystery. What is the orientation of the mineral? We have seen something of a tendency of the horizontal in the phenomena of layering. But that quintessential representative of the mineral realm, the crystal, defies a simple orientation. It grows in all directions. We have tentatively suggested the parallel lines that connect to the plane at infinity. Can we think of the orientation of the mineral as stretching from the local point to the infinite surround?

The sequence from mineral to human shows a radical reorientation. Was the upright human stance inherent in the very first place? In his book, *Developmental Dynamics in Humans and Other Primates,* Jos Verhulst suggests this could be so if each is a transformation of the other, albeit a hugely radical one between each kingdom. Figure 6.26 shows Verhulst's main line of human embryonic development which reflects evolution. The side branches lead to the animal, which is off the direct line to the human. Was there a slow and punctuated revelation of the kingdoms? The question raises many possibilities, and is left open here.

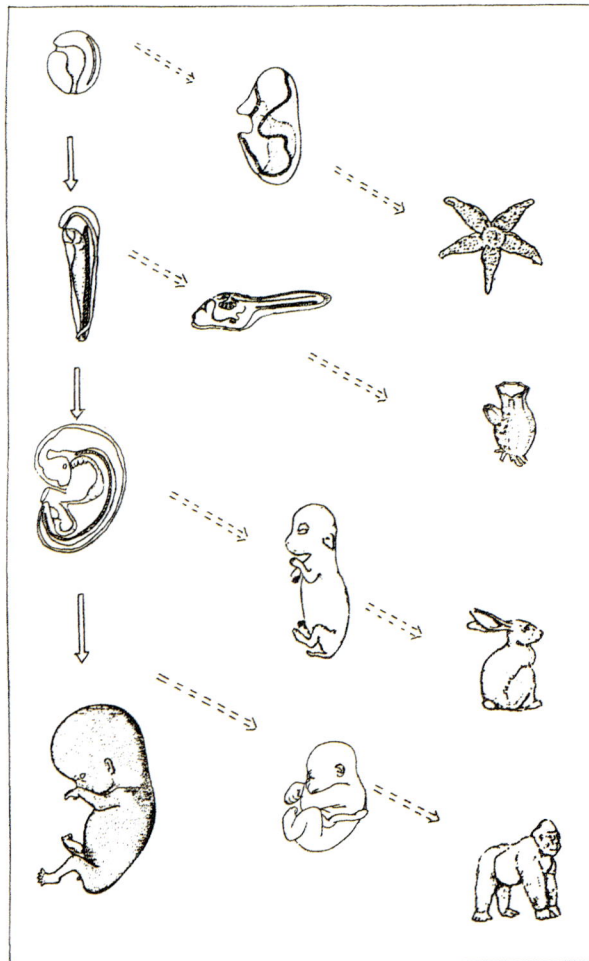

*Figure 6.26 Human embryonic development as a reflection of evolution, with side branches* [*Verhulst,* Developmental Dynamics, *p. 341*]

7.1

7.2

7.3

7.4

7.5

7.6

7.7

# 7. Measures of a Line

## 7.1 Transformation of a line

The line is a central element of the geometric world. As we saw earlier each line contains an infinitude of points and an infinitude of planes. Just as earlier with Desargues Triangle Theorem we saw that a triangle can be projected into another one, we can project a line into another. But can we also take the case of projecting a line onto itself? What changes then? The line itself remains the same, but the points and the planes within it transform.

If we do this, and watch what happens to the points, we find they all transform into another one, except for two points. These two simply transform into themselves; they remain stationary. They are called the *invariant points.* We can find these invariant points by following through the steps of this transformation (Figures 7.1 to 7.11).

Select a line (Figure 7.1), and select any point on this line (Figure 7.2, black point).

Now put any line through this point (Figure 7.3, red line), and select a point on the (red) line (Figure 7.4, red point). The black point becomes the red point after *translation.*

Rotate the red line through an angle (Figure 7.5, green line). This green line cuts the black line in a *different* place, giving a new point on the original line. (Figure 7.6, green point). One can also see this as the red point *translating* to the black line.

So the initial black point has been transformed into the new green point on the original line. In the process we have translated, rotated and then translated again. Now we let the new green line transform into another line by randomly rotating in the new green point (Figure 7.7, second red line). The two red

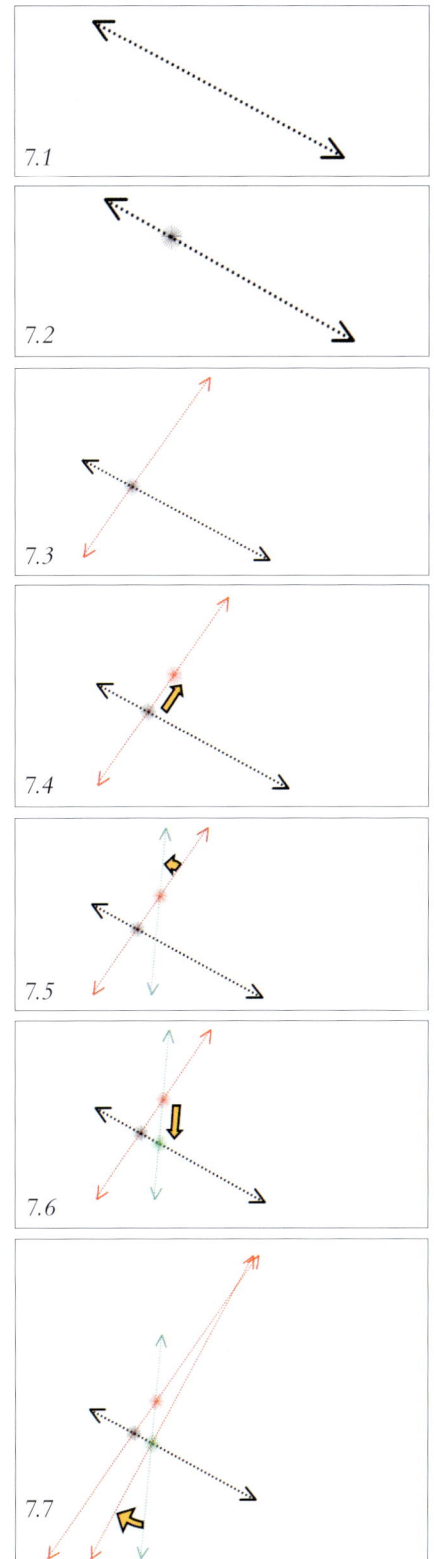

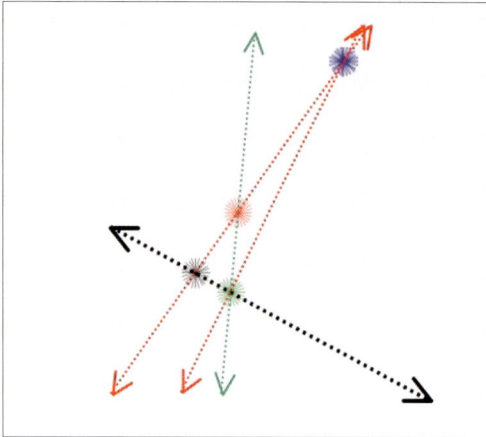

Figure 7.8

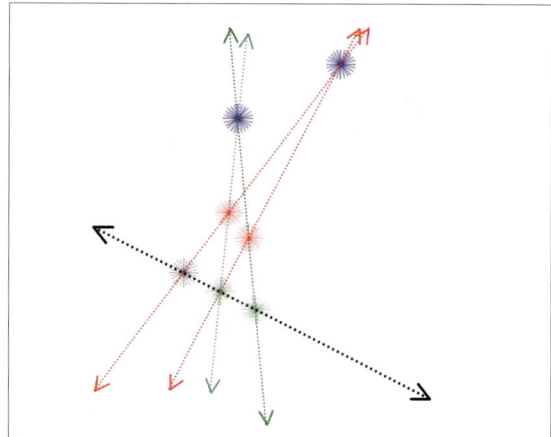

Figure 7.9

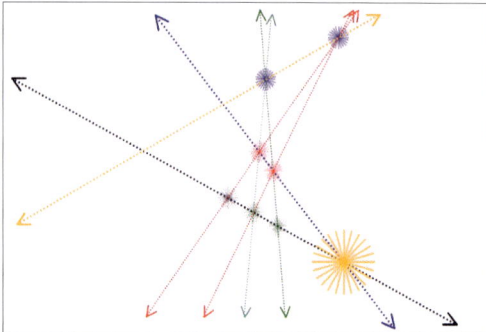

Figure 7.10

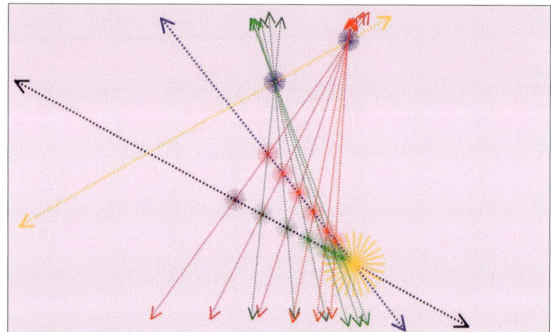

Figure 7.11

lines now cross in a point (Figure 7.8, blue point). This point is important as all the red lines will rotate through it.

We now select a second point on the existing green line, through which all the green lines can rotate. For convenience, choose a position so that the main part of the diagram fits easily on the page. Next, rotate the first green line in the second blue point to become the second green line (Figure 7.9, second blue point and second green line). This creates the second green point on the original black line.

Until now, we could have chosen any point or any angle of rotation. From the moment the second blue point is chosen, everything has been decided.

Although it may not seem like it yet, the construction has revealed where the two invariant points are. These two points are determined by the crossing points on the black line of firstly the blue line (that goes through the two red points), and secondly the orange line joining the two blue points (Figure 7.10). If the green points are continued, they crowd in towards the orange point but can never reach it, getting closer and closer all the time, as do the red points too (Figure 7.11). It is no different going in the other direction (towards

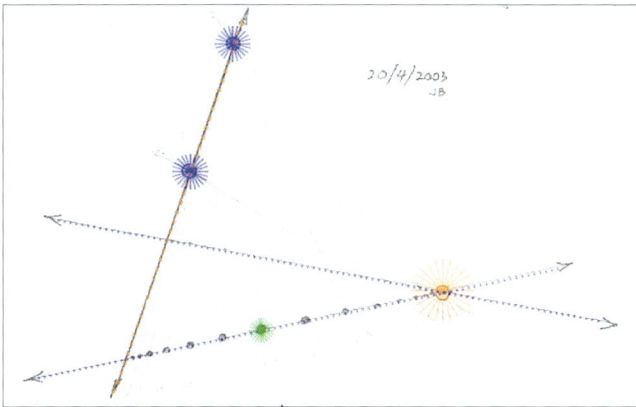

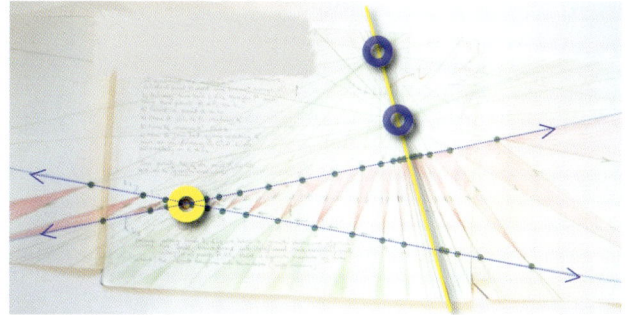

the orange line) from the original black point. The points crowd towards it but never reaches it. But note the qualitative difference: one invariant point is generated by two lines crossing, while the other is generated by the join of two points – classic dualism.

## 7.2 Growth measure

This sequence of points on a line are easy to draw. Rather than go through the apparently convoluted process above, we can simply draw our starting line, and choose the (orange) point and the (orange) line. Then select the two (blue) lines (through the orange point) and the two (blue) points (on the orange line). Now select any starting point (green) one blue line, and join lines to the two blue points. The rest follows (Figure 7.12).

The series of green points on the base line that are generated by traversing to and fro between the blue lines using the blue points, is a sequence that is sometimes called a *growth measure*. The more technical books call this *hyperbolic measure*.

We have only drawn the points that lie between the orange point and the orange line, but the sequence continues outside these two for a line is a continuous entity. These points are constructed in precisely the same way as the inner points (Figure 7.13). The entire line is a single sequence of moving points except for the two invariants. Additionally in each of the two blue points there are rays of lines reflecting the growth measure in their angles. So in the one basic transformation drawing there are four measures: two of points and two of lines.

*Figure 7.12 Growth measure*
*Figure 7.13 Growth measure beyond the invariant points*

*Figure 7.14 An exploratory drawing of bamboo nodes*
*Figure 7.15 Palm*
*Figure 7.16 Jacaranda*

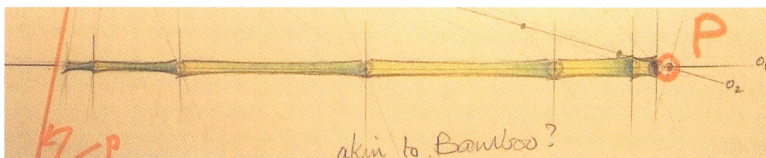

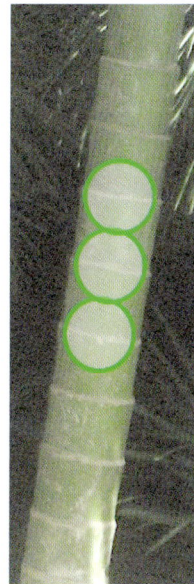

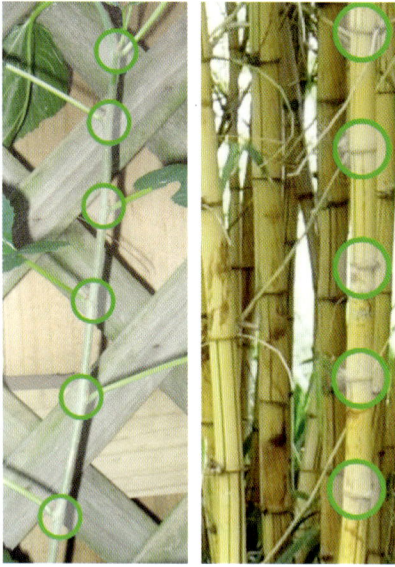

Figure 7.17 Mulberry
Figure 7.18 Bamboo
Figure 7.19 The internodal space increasing
above ground level in a palm tree
Figure 7.20 Bamboo internodal space
increasing

Figure 7.21 Casurina nodes
Figure 7.22 Wild olive

Many years ago I wondered whether these patterns might be similar to those in bamboo stems. Since then I have looked at many other plants. The distances between nodes may be short or long compared to stem length. Some nodes are close together (like Casurina shrubs), others are very far apart (some grasses).

The interesting thing (from the geometrical view) is that the node spacing towards the base of the stem, or sometimes even towards the root of a branch near the main stem, often *decreases*. We see this in palms and bamboo (Figures 7.19 and 7.20). In some cases the nodes at the upper ends of stems or branches also come closer together. For instance the fine casurina 'leaves' if you look closely (Figure 7.21) or a wild olive stem (Figure 7.22) where the nodes – accompanied by their own pairs of leaves – distinctly come closer together the further we go up the stem. A powerful example of this trend, is the growing point of a bamboo shoot I slit open for examination (Figure 7.23). Here the nodes get closer and closer, seemingly endlessly, almost looking as if they were approaching a point at infinity!

These examples point to the nodal rhythm of some plants appearing to mirror growth measure. Often at both ends of the stem the nodes close up. In the middle the distances

between nodes are larger. While we have not analysed these nodes to see whether they are true growth measures, their appearance nevertheless suggests something of the quality of growth measure. In the last example we might even be led to ask whether the actual live growing point is like an invariant point.

The geometry demands a series of further points or nodes beyond these two invariant points. Can we find anything reflecting this? Below ground there may be some hints of rhythms in root structures, but this is a lot harder to study, and I have not done this. I did, however, notice an intriguing detail with the bamboo. The root nature seems to sneak up into the leaf stem: it is as if the root tendency were trying to rise above ground, but with progressively less expression as we ascend (Figure 7.24).

At the other end of the stem, in the light, where flower and fruit preside, something quite other may be at work. Does the flower and fruit nature (for instance in colouring) intrude *downwards* into the leaf and stem nature in any way, even rhythmically, as the roots seem to work? An example of this tendency is the oleander flower and its adjacent stem (Figure 7.25).

*Figure 7.23 Bamboo growing point*
*Figure 7.24 The lower nodes of a bamboo shoot express something of the root nature*
*Figure 7.25 The colouring appears to creep down the stem from the oleander blossom*

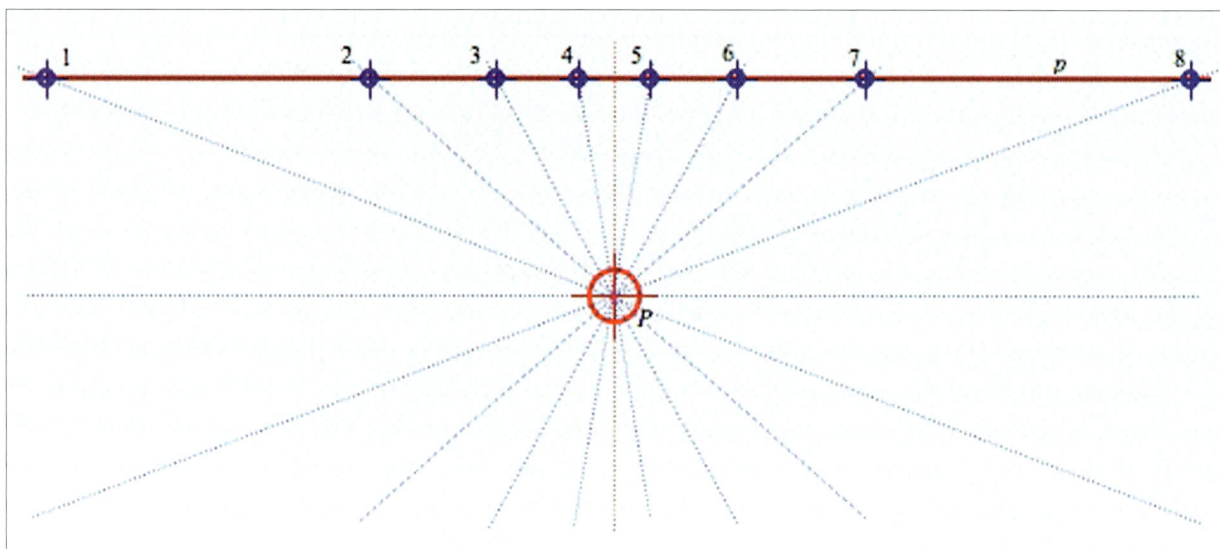

*Figure 7.26 The points of a circling measure*

*Figure 7.27 A sheaf of planes in a line*

## 7.3 *Circling measure and step measure*

Growth measure is not the only measure of points in the line. There are two others: *step measure* (formally called parabolic measure) and *circling measure* (or elliptic measure). Earlier (in Chapter 5) we used a *circling measure* as part of the construction of an *asymmetric* symmetry. Here a pencil of lines around a point is intersected by a line. The intersect points travel along the line in an ordered and harmonious fashion (Figure 7.26). These points are never still, and there are no invariant points.

Step measure, or parabolic measure, will be picked up later in section 10.3. Step measure is a special case where the two invariant points converge into a double point, and the intervals between points can become equal after a further projection.

So in any line there can be three kinds of measure:

Growth measure (or hyperbolic measure) where there are
     two fixed points.
Step measure (or parabolic measure) where these two points
     converge to a double point.
Circling measure (or elliptic measure) where the two points
     are in motion.

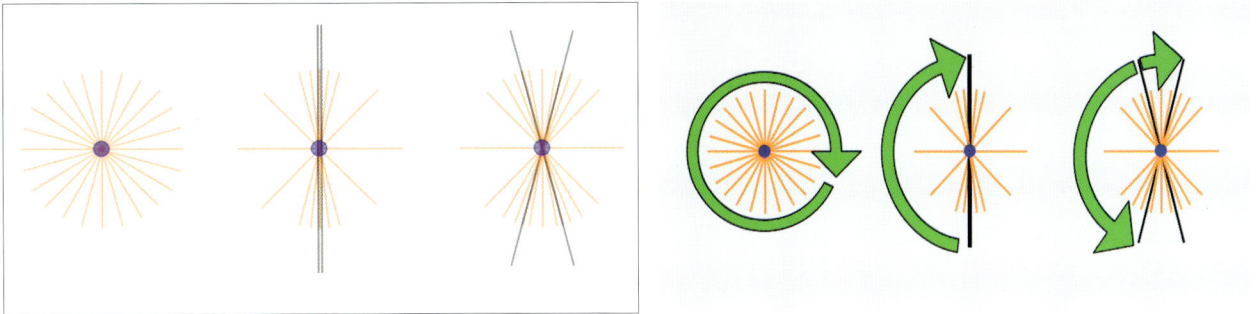

## 7.4 The planes of a line

Just as there are three kinds of measures of *points* in a line, there are also three measures of *planes* in such a line. One has to imagine a fan or sheaf of planes in a line (Figure 7.27), and of course each plane is infinite in extent. This means that any one of the planes in its rotation sweeps around the entire realm of space.

We can show the three measures of planes schematically by viewing the line end on, so that it appears as a point. In Figure 7.28 the left hand figure indicates a sheaf (or fan) of planes rotating in circling measure; every plane is in movement and there is no invariant plane. The central figure indicates a step measure where there is a double invariant plane. The right hand figure indicates a growth measure of planes with two distinct invariant planes. Figure 7.29 indicates the motions about this core line, the black lines representing the fixed planes. Obviously the direction of rotation can be reversed.

*Figure 7.28 Left to right: Circling measure, step measure and growth measure in the planes of a line*
*Figure 7.29 The direction of movement of the planes*

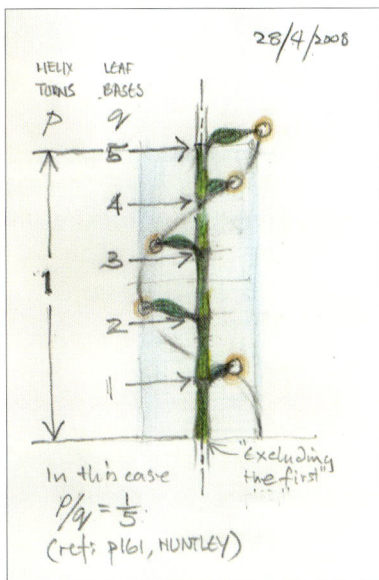

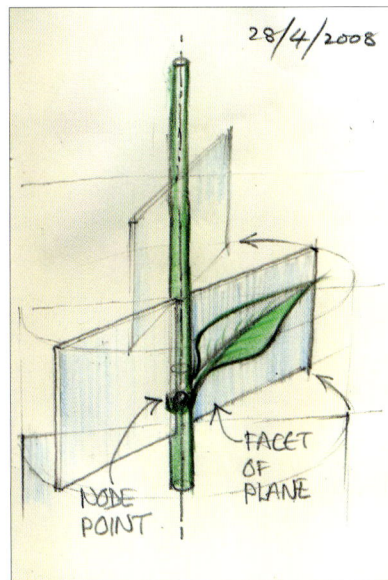

*Figure 7.30 Phyllotaxis (after Huntley)*
*Figure 7.31 The plane of stem and leaf (or branch)*

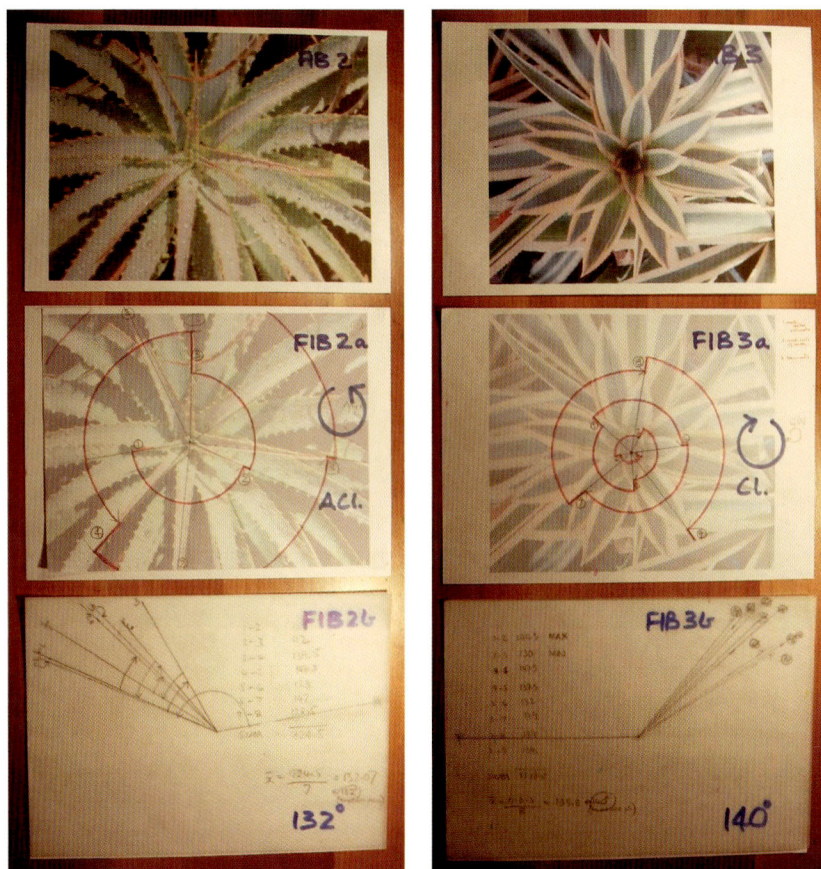

*Figures 7.32 to 7.34 Measuring and calculating the angles of the leaves growing from the stem of various plants. Note that all three come close to the golden angle φ (phi), which is about 137.5°*

Figure 7.32

Figure 7.33

We can find something of this in plant forms. There appears to be a circling measure of planes in the stem. Branches or leaves extending from the stem rotate about it in an ordered fashion. In a favourite book of mine, H.E. Huntley's *The Divine Proportion*, there is a diagram that describes this rotational order. I have redrawn it in Figure 7.30. Going up a stem Huntley describes the number of leaf bases, *q*, before the one directly above the initial one passed to the number of turns to get there, *p*, (not counting the first) as a fraction, *p/q*, which is 'characteristic of the plant' and tends to be a Fibonacci number. This turning is called *phyllotaxis* and is well known, although in most school books on botany it is not given much space. Yet it is fundamental to the plant form. It is also mentioned in several other books, particularly in the early twentieth century by authors such as D'Arcy Wentworth Thompson, Sir Andreas Cook, Samuel Colman and A.H. Church.

We are, of course, simplifying here in that we regard the plane to be the common plane between the (vertical) stem and the branch or leaf (Figure 7.31), just as we regard the node to be a point.

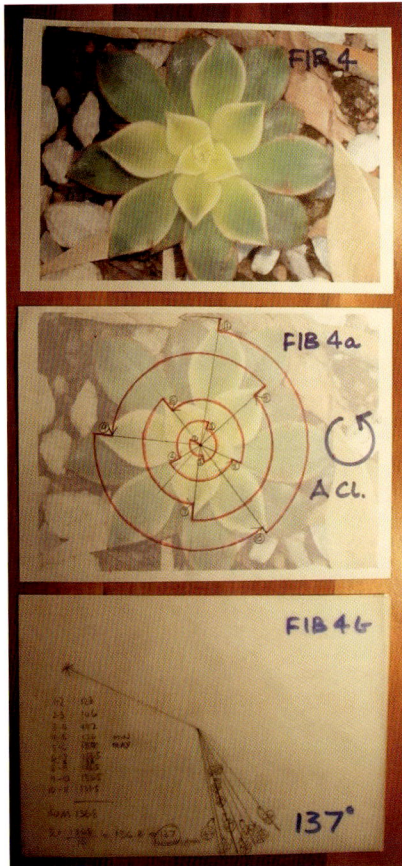

Figure 7.34

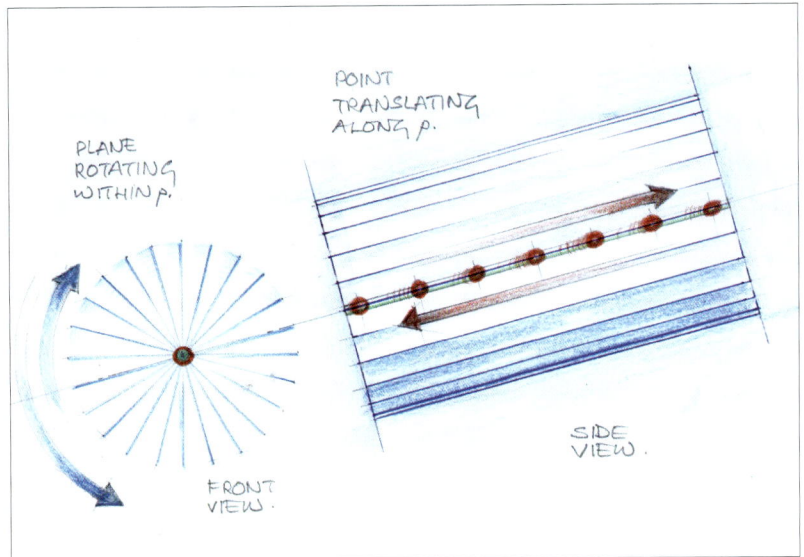

*Figure 7.35 A crude depiction of the transformations of point and plane in a line*

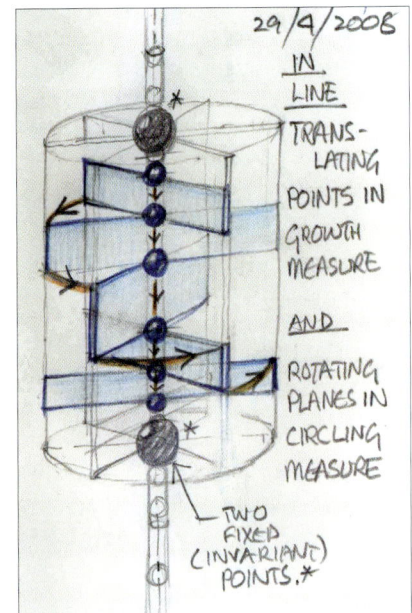

*Figure 7.36 Points moving in growth measure and planes in rotating circling measure*

Some time ago I explored the successive angles at which the leaves grow from the stem of three plants. Unfortunately I don't know their names, though they seem to be some kinds of succulents.

Peter S. Stevens describes a calculation for the golden angle, *phi,* and its relation to the golden section in his *Patterns in Nature,* and shows a number of spiral graphical layouts similar in format to plants I have shown. He also describes a spiral formed from progressive steps of this angle around a centre and then how this relates to pairs of contra rotating spirals each pair having their own relation to the Fibonacci numbers.

We can conclude that in the plant both the point and the plane are active in the line. As the nodes translate up (or down) the stem the leaves or branches rotate in their planes around the stem. Both these qualitatively different motions are going on at the same time.

An unexplored realm is how these transformations relate to the animal and human world. There may be something similar in the spacing of the 'nodes' of the spinal column.

# 8. Spirals in Nature

The spiral is a beautiful form and found in an astonishing variety in nature. Many books about forms in nature use the nautilus or some other spiral on its cover. There are a number of kinds of spiral including the Archimedean spiral, the equiangular, or logarithmic, spiral and the hyperbolic spiral. We shall begin with the simplest.

## 8.1 Archimedean spiral

This spiral has the simple property that for equal amounts of angle rotated, the radius increases in equal steps. This is more easily grasped if we call it the rope spiral.

This kind of spiral does not appear much in nature. At a cursory glance some little tube-like shells like *Olivancillaria acuminata* appear to be Archimedean spirals, but even they get a bit larger as they grow, as can be seen simply by looking at the centre.

*Figure 8.1 A coiled rope illustrating the Archimedean spiral*
*Figure 8.2* Olivancillaria acuminata

## 8.2 The equiangular or logarithmic spiral

The equiangular, or logarithmic, spiral moves out from the centre with ever increasing rapidity, quite unlike the Archimedean version. I would love to call it a Bernoulli spiral after the Swiss mathematician Jacob Bernoulli (1654–1705) who called it *Spira mirabilis,* the 'marvellous spiral'.

This form is fundamental in a huge number of nature's artefacts and phenomena. It is probably safe to say that *all* plants exhibit it in some way, however obscure. Countless seashells and land snails exhibit it. Sunflowers, whirlpools, the spiralling clouds of cyclones and the great galaxies exhibit it. It even crops up unexpectedly, for instance in the way hawks approach their prey, as their sharpest view is at an angle to their direction of flight.

How does this spiral behave? Why equiangular? For every equal angle turned the spiral always maintains the same slope or angle to a radius line drawn from the centre (Figure 8.7). Hence it is called an *equi*angular spiral. This means that all such tangents around the spiral are at the same angle to their radius line (hence the hawk's flight). So, from a point centre, the spiral form, as a curve in the plane, strives ever outward towards the line at infinity. But at the same time, the geometrical construction inwards will never reach the centre.

## 8.3 The general spiral

The equiangular spiral is actually a case of a far more general curve. A general spiral, in a plane, I would like to call a *spiroid,* a contrived term, but it fits the case. The general spiral can be constructed out of the *activity* of the two geometric elements in the plane and their

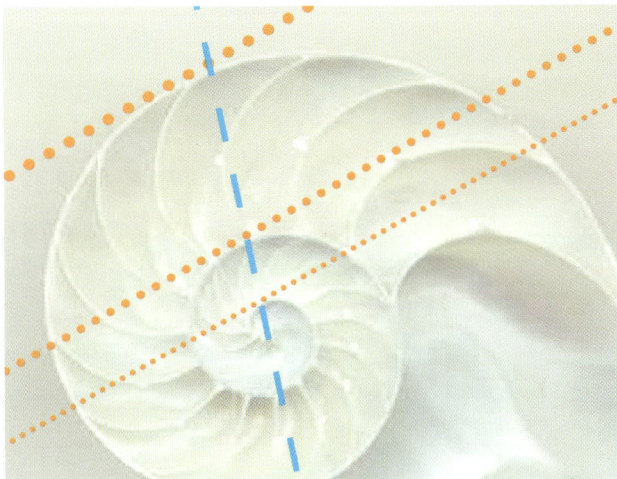

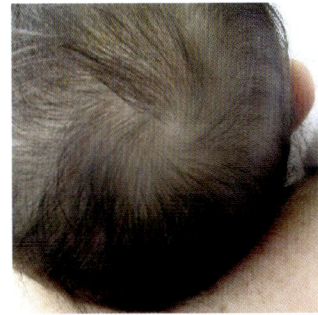

*Figures 8.3–8.6 Equiangular spirals in nature. A hair whorl, land snails from Tunisia, an ammonite fossil, or the spikes around a plant*

*Figure 8.7 Section of a chambered nautilus. Note the blue dashed line through the centre and the three orange dotted tangents. They are all parallel and at the same angle to the blue dotted line*

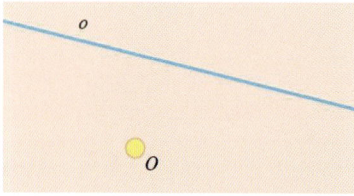

Figure 8.8

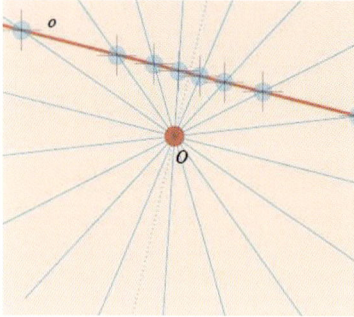

Figure 8.9

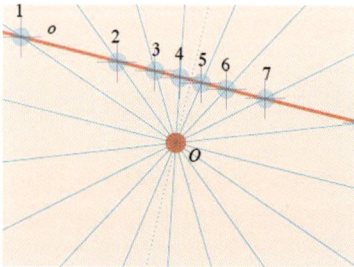

Figure 8.10

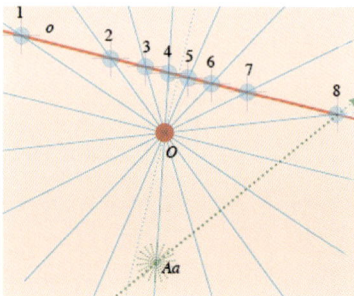

Figure 8.11

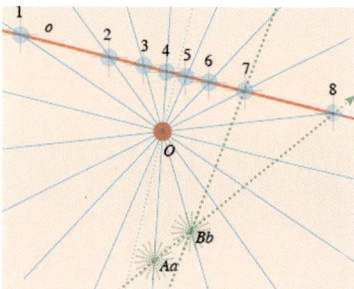

Figure 8.12

intrinsic behaviour. These are point and line, represented by the yellow circle, O, and the blue line segment, o, in Figure 8.8 (Point and line).

The next step is to select a family of lines in O and a family of points in o where both of these families are linked. We put a series of lines in point O such that they are all (say) 20° apart creating a pencil of lines, or what some might call a star (Figure 8.9). This pencil of lines cannot help but interact with the line o, thus creating a series of points. Each blue line and its blue counterpart point are named in order from left 1, 2, 3, and so on. They are called point/line pairs (Figure 8.10). The point/line pairs 1 to 7 are visible, 8 is at the edge of the figure, and 9 is at infinity.

With this structure we have created a field or framework. Now we insert a point/line pair Aa into this mini field, and see how it has to move (Figure 8.11). Point A happens to be on line 4 and line a happens to be in point 8, but this is random. Now we make Aa move in the field. Point A can only move or translate in a line, and line a can only swing or rotate in a point. This means that A translates to a point on another line, either 3 or 5, let us take 3, and it becomes B. Then line a turns (or rotates) in B and becomes line b where it now meets point 7 (Figure 8.12).

Then this process is repeated. This leads to Cc, Dd, Ee and so on, and a step-wise curve arises. Or one could say that all the transformations of the point/line pair lie on a curve. This curve is called a path curve. (Felix Klein and Sophus Lie discovered and published these in the nineteenth century. Called W-Kurve in German, they were misinterpreted as Weg-Kurve and translated into English as path curve.)

Figure 8.13 The complicated spiral

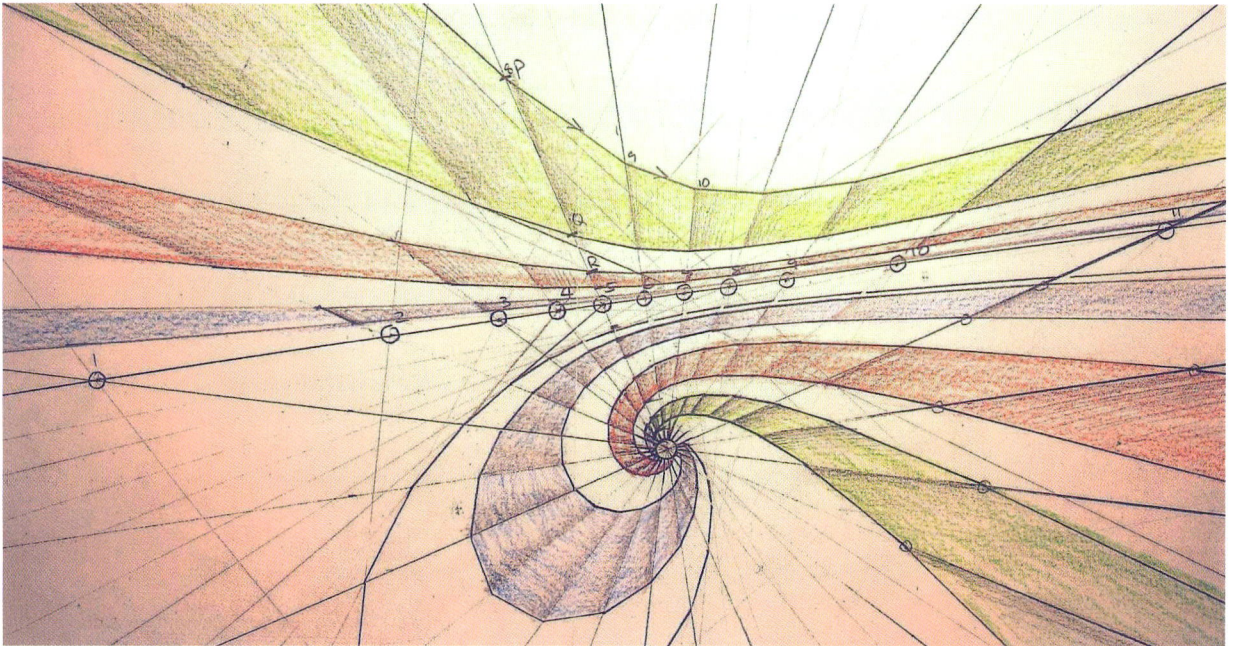

*Figure 8.14 Further spirals from the same field*

The curve that arises from the above construction is what I call a *spiroid,* as the tendency is for the point/line pair to curve towards the centre point *O.* In fact, like the equiangular spiral, it keeps going round *O* never quite reaching it (Figure 8.13). Two curves are actually shown. This is the beginning of a whole field that covers the entire plane. Figure 8.14 shows how the field exists quite happily on *both* sides of the line *o,* the curves having gone through infinity to return from the other side. Note that the spirals constructed here are asymmetric.

## 8.4 The regular equiangular spiral

In nature we usually find regular spirals. To construct such a regular spiral is fairly easy. All we have to do is to move the line, *o,* to infinity while keeping point *O* central. To draw it, it is best to use an A3 (or double letter $18 \times 11$ in) sheet of paper, a pair of compasses, a long ruler, a protractor and a sharp pencil.

Mark a point, *O,* in the centre of the page, and through this draw a horizontal line, (Figure 8.15). Starting from the horizontal, using a protractor mark every 20° around point *O* (Figure 8.16).

Now add a green point/line pair *Aa.* The point *A* must be on one of the blue lines (say line 2) and the line *a* must be *parallel* to one of the blue lines (say line 4) as in Figure 8.17. It must be parallel because the line *a,* must go through a point at infinity, point 4 (on the line at infinity, *o,* suggested by the dashed red circle).

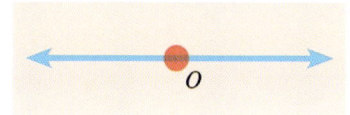

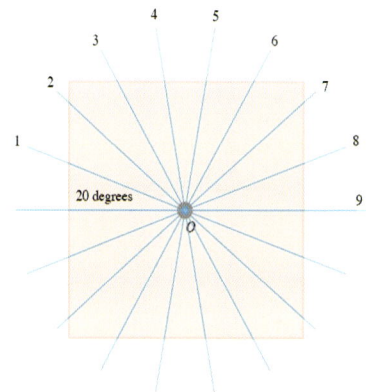

*Figure 8.15 and 8.16 Constructing a regular equiangular spiral*

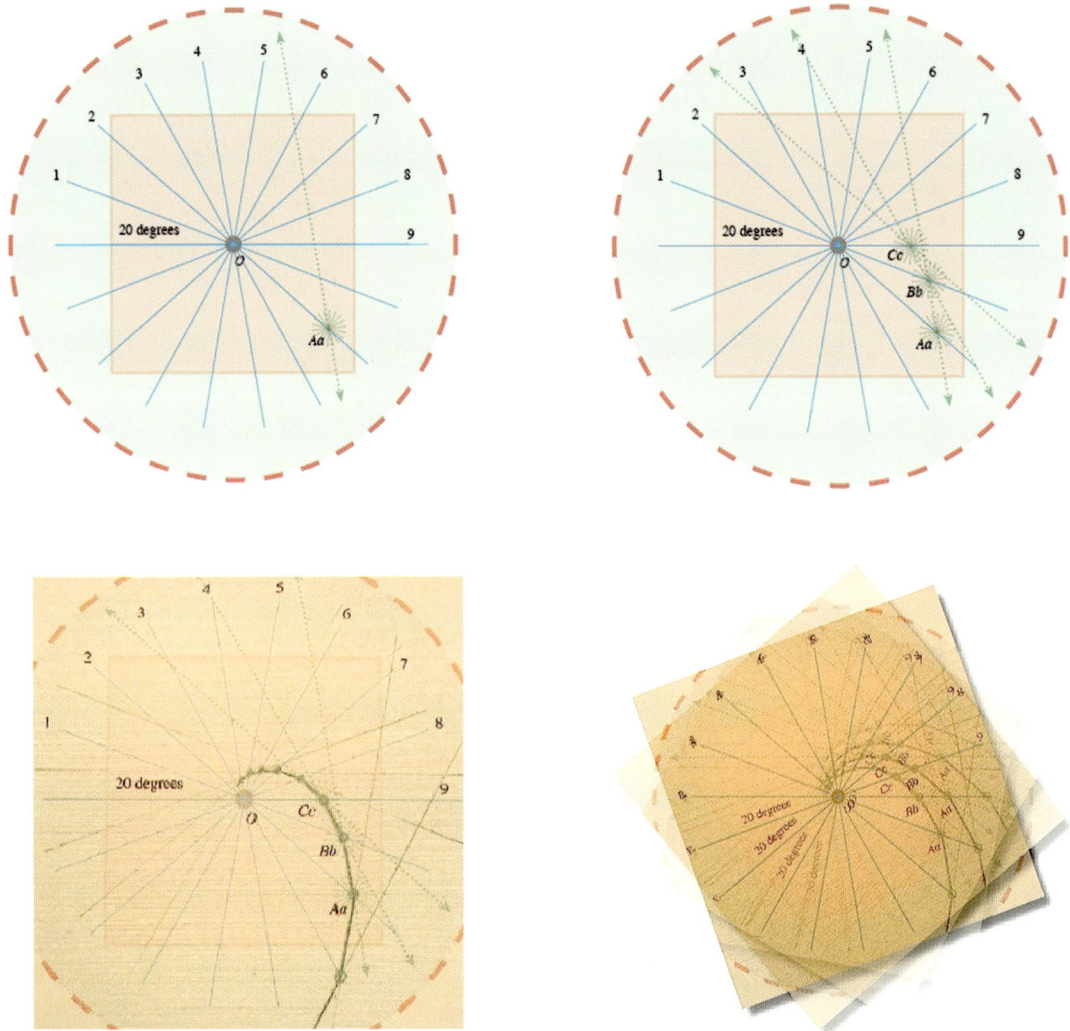

Figure 8.20 The beginnings of a field of spirals

*Figures 8.17–8.19 Constructing a regular equiangular spiral*

Let point *A* translate to *B* (on line 1), and rotate line *a* to be parallel with line 3, making line *b*. This is the point/line pair *Bb*. By repeating the process we find point/line pair *Cc* making the line parallel to 2 (Figure 8.18). Then *Dd* becomes parallel to line 1, and so on. We see a curve developing through repeated translations and rotations (Figure 8.19). We see the form of the equiangular spiral emerge from this combination of movements, winding anticlockwise into the point *O*. Or, if we reverse the process, it is winding out and out towards the line, *o*. This is going to take a long, long time!

Of course there are a great number of such spirals that can be constructed from this field. A few are indicated in Figure 8.20.

## 8.5 *Nature's spirals in the plane*

It is common to ask how life evolves from dead matter, but should we rather be asking how is dead matter the product of a living process? The laws of physics certainly pertain, but they seem to work against life as such. When an organism dies, it is given over to physical laws, and it disintegrates. What is living appears to be constantly striving for the perfection of the ideal form, despite the vagaries of the environment. Living organisms strive to geometric form. It is never quite perfect, but it nevertheless is the quintessence of an organism that it expresses *form*. For instance, a little shell might be truly dead, broken and worn, but we immediately recognise it as a shell of a one-time living creature. Its *form* reveals more than the mere calcium content. Through its inherent dynamic geometrical forms it appears to contradict the mineral laws of decay and entropy. It came about because of life, not death.

The organic form can expect countless disruptions yet so often can recover its form astonishingly well, as we can see in the broken and recovered shell of Figure 8.23. It can help to think that it is merely re-entering the inherent existing form field, which is unique to that species. It strives for this and somehow manages it remarkably well.

So, how close can the material expression marry the ideal? We shall analyse the fossil of an ammonite, an equiangular or Bernoulli spiral. This fossil is virtually a flat spiral, so it is a good candidate for this analysis.

The first step is to *assume* that an equiangular spiral of some proportions will describe the planar section of this fossil form. Every such spiral has a point as centre. The second step is to estimate the centre for this fossil. Call this centre *O*. The third step is to impose a coordinate system. Rather than the common Cartesian, *x*–*y* coordinates, we choose the polar coordinate system because it deals with rotation much more readily. Figure 8.25 shows the array of lines at 30° intervals. Next we mark in some data points on the radial lines where they cross the spiral profile. This makes twelve data points per circuit. These points too will inevitably be estimates.

I now use the technique of section 8.4 to plot an estimated spiral. This develops the motion of point/line pair to create the regular equiangular spiral, as we can see from Figure 8.26 (the same figure used previously). Select a point line pair *Aa*. From estimates I guessed that the line should be about

*Figure 8.21 A dead shell*
*Figure 8.22 A whole shell*
*Figure 8.23 A broken and recovered shell*

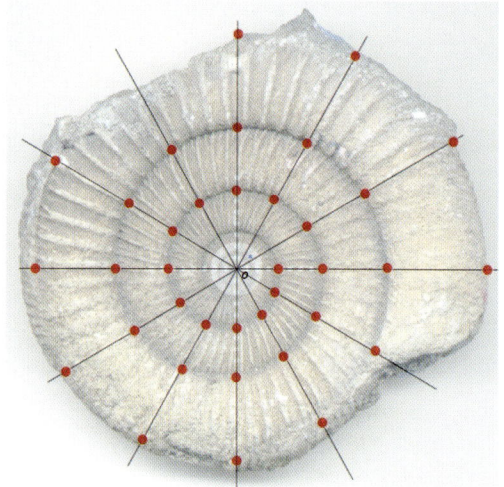

*Figure 8.24 Fossil ammonite*
*Figure 8.25 Superimposed polar coordinates and data points*
*Figure 8.26 Construction of a regular equiangular spiral*
*Figure 8.27 Estimating the angle of the spiral*

*Below:*
*Figure 8.28 Radius values calculated with a simple computer program*

```
r =   91.7
r =   87.928359
r =   84.311846
r =   80.844081
r =   77.518947
r =   74.330575
r =   71.273342
r =   68.341854
r =   65.530939
r =   62.835637
r =   60.251193
r =   57.773049
r =   55.39683
r =   53.118347
r =   50.933577
r =   48.838668
r =   46.829923
r =   44.903798
r =   43.056895
r =   41.285955
r =   39.587855
r =   37.959598
r =   36.398311
r =   34.901241
r =   33.465745
r =   32.089292
r =   30.769452
r =   29.503898
r =   28.290396
r =   27.126806
r =   26.011074
```

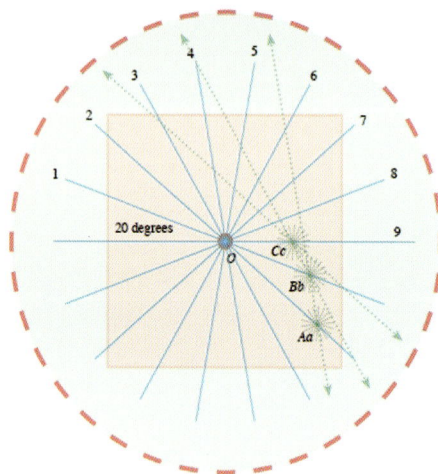

79 degrees to the radiant lines from *O*. Hence starting at *A*, I made the point/line pair swing through an angle of 79° at every 30°radiant. Continued around for half a cycle we come close to six of the chosen data points. This is a good indicator that the angle selected is not too far out. With this as an estimate it is then not difficult to test the actual spiral profile against the geometric construction that gives an equiangular spiral.

To plot the spiral I wrote a simple program. It is easy to see where and how the data points (red dots) relate to the calculated points (red circles) in Figure 8.29. As we go towards the centre the data points start to wind in a little more rapidly. Generally it is a fairly good match, considering all the estimates made. Possibly the broken outside of the fossil was not the true position of the spiral, and there may have been a better fit starting a little

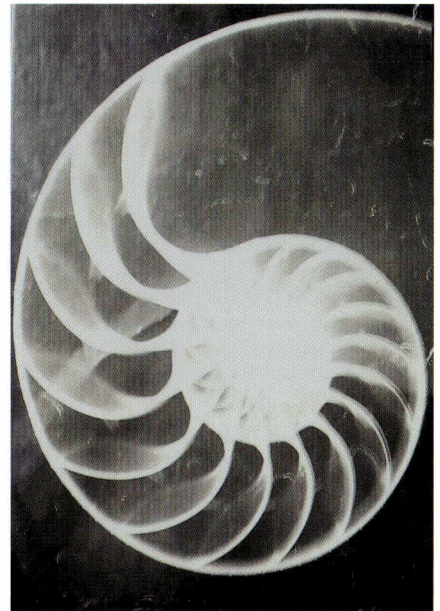

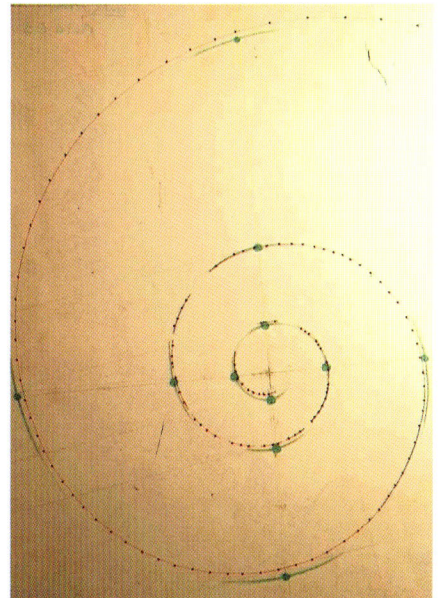

further in. However, this initial exercise suggested that this kind of approach was worth pursuing with other artefacts.

The other example I looked at some time ago was the famous nautilus. Only the result is shown here from an X-ray published on an *Australian Academy of Science* poster Figure 8.30 (Nautilus section). Again the data points are estimates and there is no certainty that the X-ray would have been taken with the shell precisely square to the viewpoint, nor that the section was precisely through the middle of the shell. Nevertheless it seemed pretty good to me and well worth analysing.

The data points were pricked through the poster and marked in red on the back (Figure 8.31). I then estimated a centre, and calculated a few points. These calculated spots are shown in green with freehand suggestion as to its further path, through such points. It is a pretty close fit. This has probably been done a thousand times before, but I think it is good to see some empirical results to confirm the match.

A third example is *Busycon perversum*. This shell is actually conical, so I examined a picture plane projection; that is, perpendicular to the long axis. Again a fairly close match between actual and calculated points is shown in Figure 8.35.

*Figure 8.29 The match of actual data points (red dots) with calculated positions*
*Figure 8.30 X-ray section of a nautilus (from an Australian Academy of Science poster)*
*Figure 8.31 Data points (small red), calculated positions and curve (green) of the nautilus shell. (As the points were on the back of the poster, the result is a mirror-image or reversed)*

*Figures 8.32 and 8.33 Busycon perversum, side view and end view*

Note the calculated spiral seems to follow round the inside of the protrusions that are almost covered by each subsequent turn. Observant readers will have noticed the direction of the spiral is anticlockwise. Is this usual for seashells? Observe for yourself.

Mathematicians have been fascinated by spirals in nature for a long time. In the early twentieth century there was a profusion of books on the subject, for example Andreas Cook's *The Curves of Life*; D'Arcy Wentworth Thompson's *On Growth and Form*; J.B. Pettigrew *Design in Nature,* and Samuel Colman's *Nature's Harmonic Unity.*

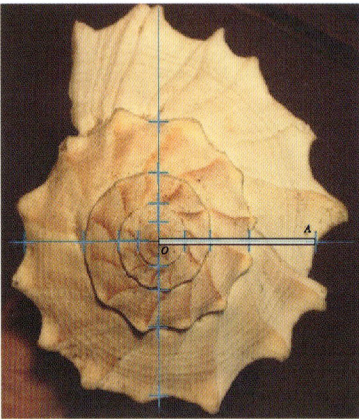

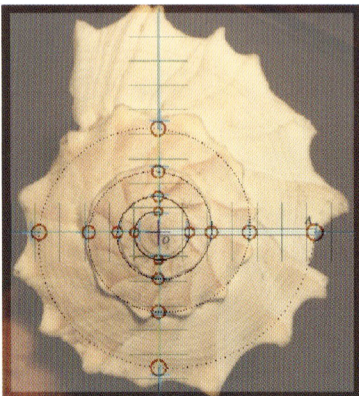

*Figure 8.34 Marking the centre, the initial axis and data points*
*Figure 8.35 A reasonably close match between actual points (blue lines and calculated points (red circles)*

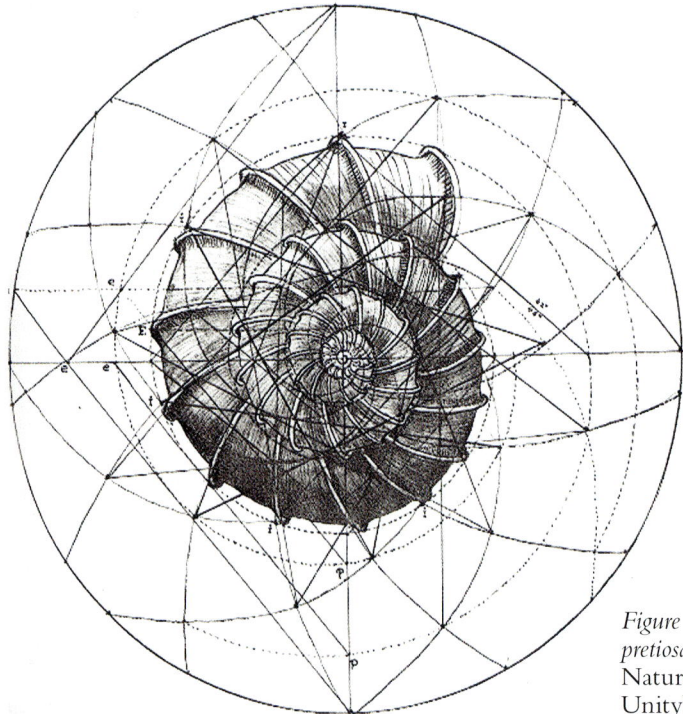

*Figure 8.36 Scaleria pretiosa [ from Colman, Nature's Harmonic Unity]*

## 8.6 Two dimensions in a point

All the geometry in this chapter has so far been in the flat, infinite plane; in other words, in two dimensions. From the point of view of projective geometry we have the elements of point and line in the plane. We can look at the polar image (as we saw in section 2.7) and look at the elements of plane and line in a point. This is another kind of two dimensionality. We find there a whole world of shapes all built of planes and lines. It is difficult to imagine, and even more difficult to illustrate, but I have attempted a suggestion in Figure 8.37. A further example of this kind of two dimensionality is shown in Figure 8.38.

The simplest figure or shape in a plane is the triangle. There is also a kind of triangle in a point. Figure 8.39 gives some idea of what it looks like. Notice our common little three-point-and-line triangle in the blue plane. This we are used to. The pyramid-like figure crossing space is less familiar as a 'triangle'. Our point-wise triangle is actually the blue point $P$ and the three planes and lines in it, and of course it extends on both sides of the point indefinitely. In order for us to 'see' the unfamiliar figure we have taken a section (where we seen the familiar triangle). Using sections to reveal the point-wise structures is a useful device employed a few times below.

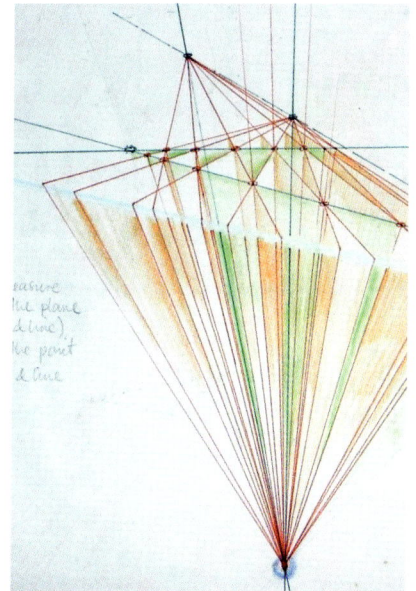

*Figure 8.37*

*Figure 8.38*

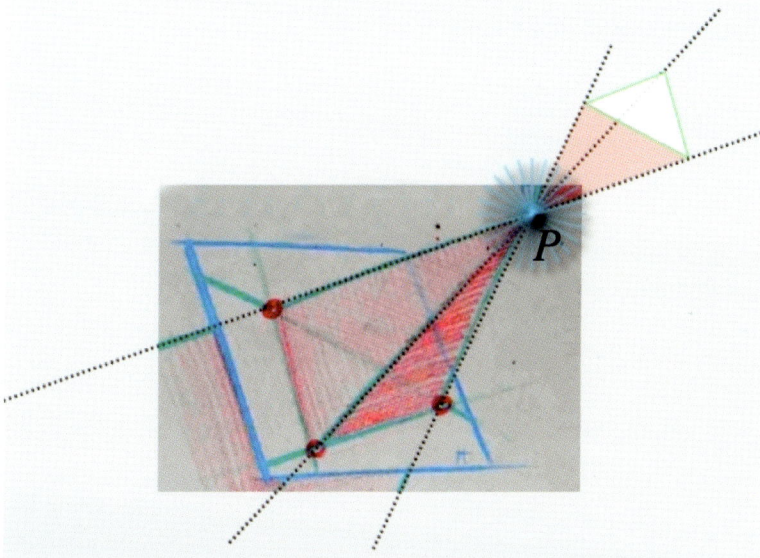

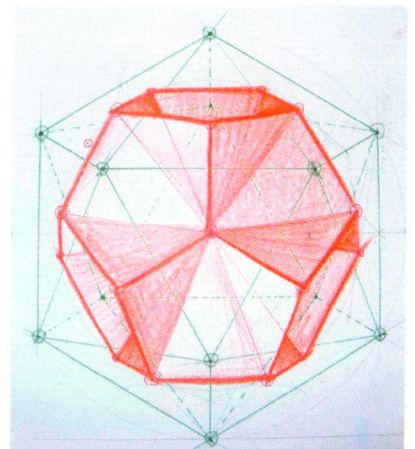

*Figure 8.40*

*Figure 8.37 A growth measure of planes and lines in a common a point*
*Figure 8.38 A hexagonal net of planes and lines radiating from a single point*
*Figure 8.39 The triangle in a point*
*Figure 8.40 Dodecahedral point figure*

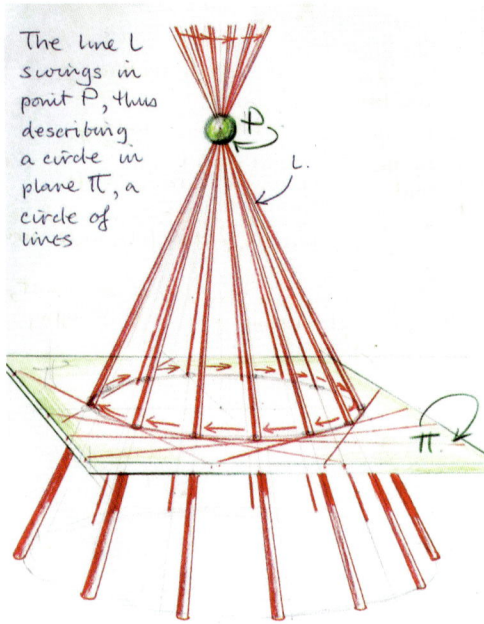

*The line L swings in point P, thus describing a circle in plane π, a circle of lines*

*Figure 8.41 A cone lines*
*Figure 8.42 Elliptical cone*

In the dodecahedral type form (Figure 8.40), there is a structure of lines and planes that describe a definite two-dimensional array through a common point. Again, to make this visible, we have made a number of cross-sections that appear as pentagonal forms. These in effect amount to stopping the raying lines at the same radius from the centre point, enabling us to see clearly what it is (the partial shading of the planar elements helps

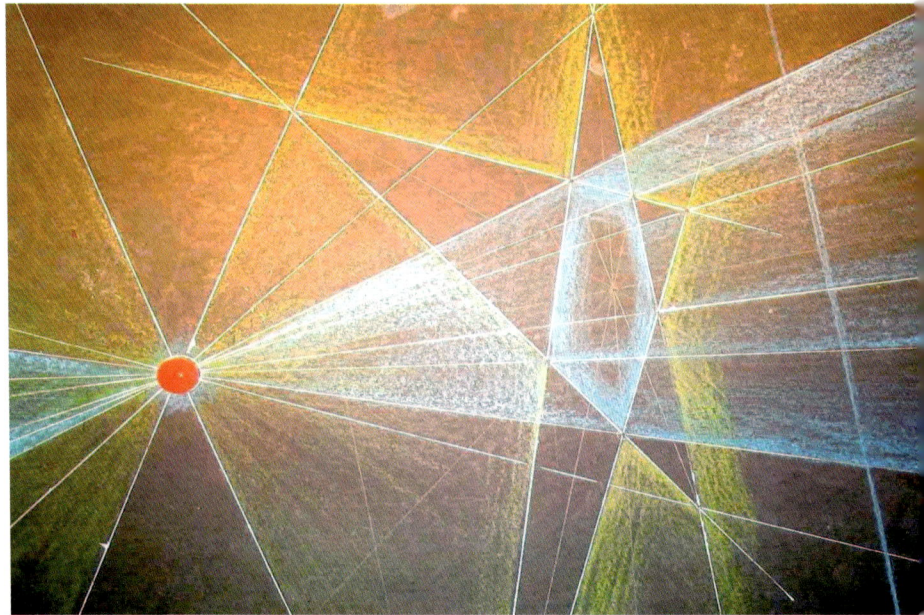

*Figure 8.43 Star in point and plane*

*Figures 8.44 and 8.45 Radiating, pointwise forms*
*Figure 8.46 Cycad fronds show a raying tendency*

too). There is a resemblance to a viral structure – a number of virus forms have an dodecahedral (or icosahedral) basis. Perhaps such tiny organisms represent the focus of some such far distant, infinite forces.

A sketch demonstrating the duality inherent in the contrast of these two kinds of two dimensions is drawn in Figure 8.41. A line, *l*, swings in point *P* and describes a circle in the plane *π* that is a section through the cone. Note that the cone extends on both sides of the point, as we saw also in the triangle form.

We can imagine an ellipse in a plane, constructed from its points or tangential lines. Similarly, we can also conceive of it as consisting of a pencil of lines or sheaf of planes through a point. Figure 8.42 shows the planar ellipse and the elliptical cone. This duality must also apply to a spiral or any other planar form.

Do we see these two-dimensional pointwise radiating lines and planes in any of nature's forms, perhaps even as a point spiral? Figures 8.44 and 8.45 suggest it, for the star appears to come from a point. The familiar dandelion clock also can evoke this interpretation. In the cases shown the 'star' may really spread or ray out from nodes slightly apart but visually appear almost coincident. The tendency is expressed in these plants, even if the geometry may not be totally accurate.

*Figure 8.47 Terebra aerolata shells placed side by side show an almost straight line, thus being almost true simple cones*
*Figure 8.48 A shell from the Trochidae family, probably Cunningham's Top, showing a good match to a true circular cone*

*Figure 8.49 A shell with slightly convex curve shows that not all seashells are true cones*
*Figure 8.50 This slightly concave profile again is an exception*

## 8.7 Nature's spirals in the point

What of spirals in a point? This is a lot harder to deal with. I have not found an easy way to make measurements of organic forms that radiate from a single point. Nevertheless, it may be worth an effort, as geometrically the point spiral is this *polarity* of any plane spiral. There is in fact a clue in the conical aspect of *Busycon* and other seashells.

From this point of view there is indubitably a point-wise cone in evidence. How close to the simple cone *are* the cone shells? Most are close, but there are exceptions. Of course strictly we are speaking of half-cones, for the shells obviously don't grow on both sides of the point of origin.

The two shells without straight line sides may still be point forms, but of another nature, possibly tending to an egg or vortex shape that will be discussed in a later chapter.

# 9. Projective Geometry in Three Dimensions

## 9.1 The simplest three-dimensional form

If we imagine the whole of three-dimensional space to be transformed onto itself by the most general one-to-one projective transformation, it can be shown that there will necessarily be just four points which will be self-transforming.* Any three points determine a plane (and a triangle), so any four points in space must determine the simplest three-dimensional form. This is in fact a tetrahedron, a triangular pyramid. A tetrahedron has four points (vertices), six lines (edges) and four planes (surfaces).

Mathematicians work with two kinds of numbers. First there are the everyday numbers we all use, including negative numbers, fractions and irrational numbers (like $\pi$) that have a never-ending string of decimals after the point. These numbers are called 'real numbers'. The other kind of number is called 'imaginary' and paradoxically is practically impossible to imagine. These are the square roots of negative numbers.

Remembering that the square of $-1$ is $+1$, it is impossible to get a square that is negative. And yet mathematicians use these numbers (denoted $i$). In projective geometry, real numbers are equivalent to real points, while imaginary numbers have a circling property. The circling measure we used earlier, is algebraically a measure that has imaginary numbers. Like the real square roots, imaginaries always come in pairs (the square root of 16 is $+4$ or $-4$, so similarly the square root of -16 is $+4i$ or $-4i$).

Returning to the tetrahedron, we find there are three basic kinds (there are also a number of special cases or degenerates).

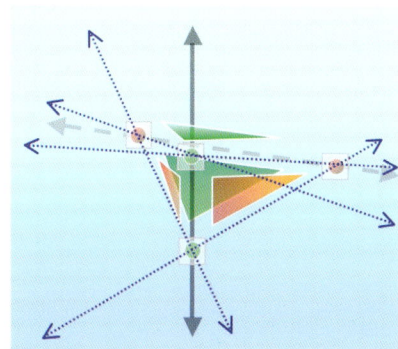

*Figure 9.1 The all-real tetrahedron*

* Lawrence Edwards shows this in *The Vortex of Life*, p. 305.

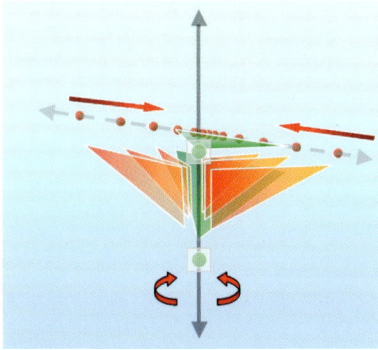

*Figure 9.2 The semi-imaginary tetrahedron*

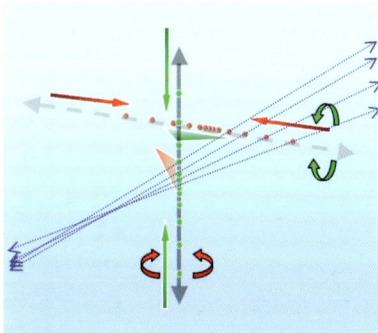

*Figure 9.3 The wholly imaginary tetrahedron*

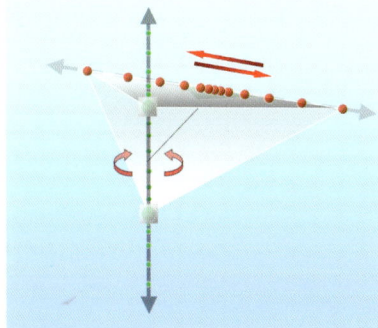

*Figure 9.4 Semi-imaginary or hybrid tetrahedron with rhythms*

First, the all-real tetrahedron (Figure 9.1). Here we have six real lines, four real points, and four real planes. I have used a square around the four invariant points as a symbol of their fixedness.

Second, the semi-imaginary tetrahedron (Figure 9.2). There are two real points, and two imaginary or circling points, two real planes and two imaginary or circling planes. This is a kind of hybrid, a mix of static and moving elements. In the figure we can see the two fixed green points in the vertical line, and the two fixed green planes in the other more horizontal line. Also indicated are the two imaginary, moving red points (circling in opposite directions) in the horizontal line and two rotating red planes in the vertical line.

Thirdly, the imaginary tetrahedron (Figure 9.3). Here, apart from two real lines everything is circling. This means that sheaves of planes are moving in both lines (clockwise and anticlockwise) and rhythms of points are moving in both lines (in both directions). There are another four lines linking the four (imaginary) points and planes; these, too are in motion, but to avoid overcrowding they are not shown in figure.

All three variants are important but the hybrid, or semi-imaginary, one especially so as we shall see later.

Let us look at one example of the hybrid type where there are the two real points in the vertical line and two circling points in the horizontal line. Making the two lines perpendicular to each other is a choice of course. And I also decide for a growth measure in one of the lines and a circling measure in the other (this is indeed possible). The *measures* inherent in these two lines, something not elaborated on before, are indicated roughly in Figure 9.4. The fixed elements are shown in black/grey (two lines, two points and two planes) and the measures of points only in the two lines, (to avoid overcrowding). The red points are an unspecified circling measure and the green points are an approximate unspecified growth measure. There is an infinite variety of these too, of course.

*Figure 9.5 Construction of a general spiral*

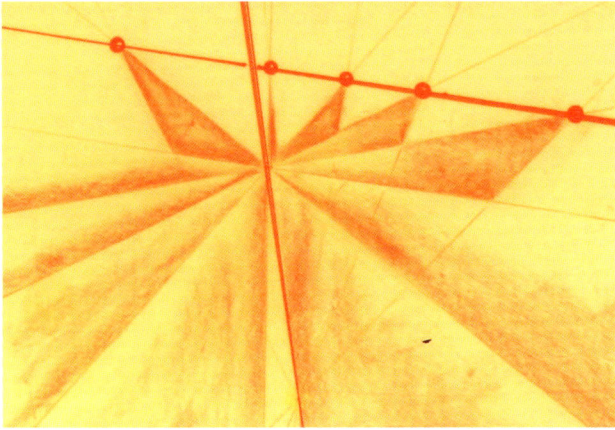

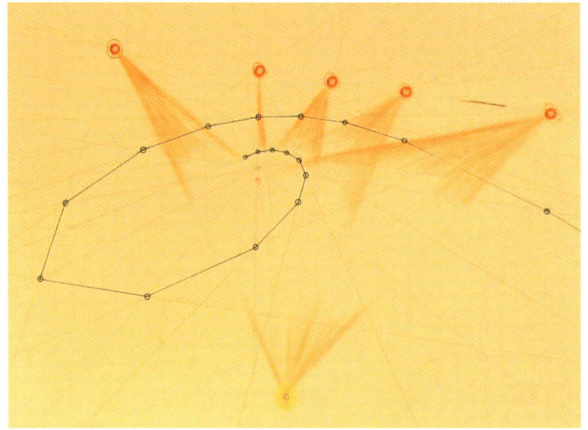

*Figures 9.6 and 9.7 Construction of a vortex-like form*

What form, or further forms, can be generated here between a point and a plane? I selected a general spiral (or spiroid) shape in the top fixed plane and a spiroid cone in the bottom fixed point. (The construction for this was elaborated in section 8.3; Figure 9.5 is a reminder.) I thought this would lead to something interesting if the two were interlocked, but it was not quite as simple as that. I found the spiroid in the plane and the spiroid in the point had to be *different*. There needed to be a displacement between the two.

To start the process the two lines (shown in orange) were drawn with an arbitrary circling measure in the horizontal line (Figure 9.6). Then in the top plane through the top point, a spiroid shape is drawn (Figure 9.7, green) as described in section 8.3, (it is the path of a point/line pair between a point and a line in a plane with measures in both point and line). Now add a conic spiroid shape in the bottom point (Figure 9.8). Now we have planar spiroid and conical point-wise spiroid, a spiroid cone, but they are different and do not intersect each other.

The next step is tricky. The two two-dimensional shapes are connected in Figure 9.9, showing how a *point/line/plane triple* moves

*Figures 9.8 and 9.9 Construction of a vortex-like form*

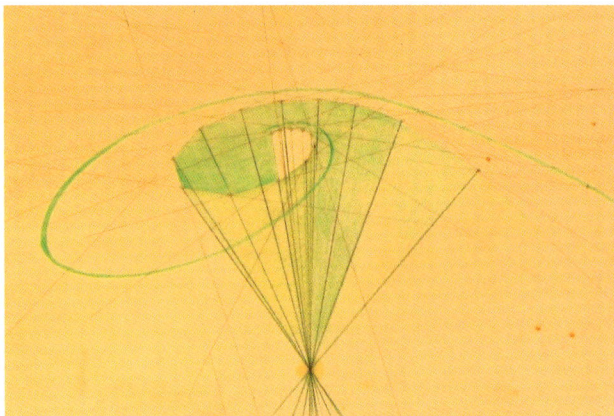

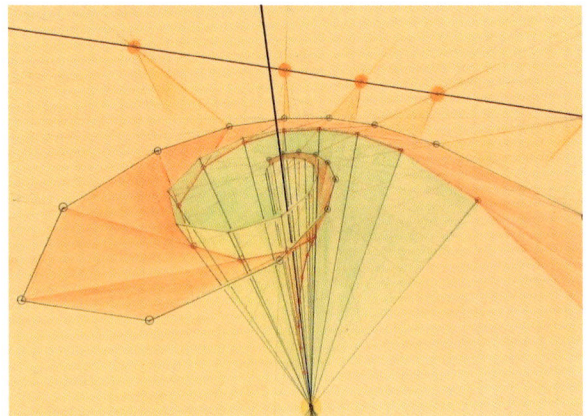

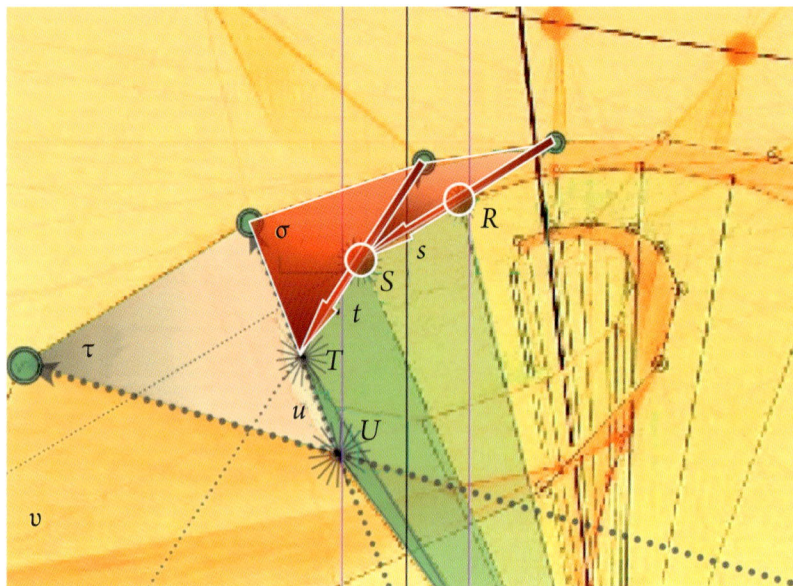

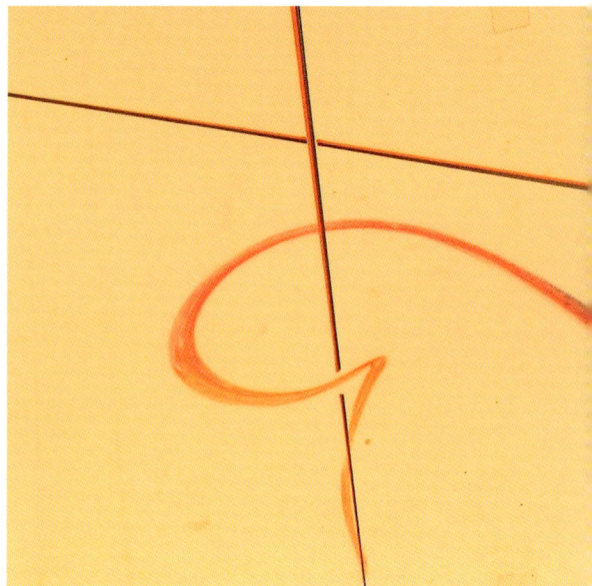

Figure 9.10 Construction of a
vortex-like form
Figure 9.11 A general three-dimensional
curve appearing as a kind of vortex

within this whole complex. A subtle curve results. Picking an arbitrary beginning point, say, point R, and allow this to *translate* along line s, to point S. Line s now *rotates* in its plane to line t. The plane now *sweeps* in line t to become plane Σ *(sigma)* (Figure 9.10). This threefold activity now repeats itself again and again, and we notice that it curls around the central vertical line moving downwards towards the bottom point.

If all the construction lines (except the two initial lines) are removed, the curve resulting from this translating-rotating-sweeping motion becomes clear (Figure 9.11). The curve is a beautiful form coming in from infinite expanses in the top plane, spinning about the vertical line and striving towards the bottom point. This three-dimensional curve is a general case for the hybrid tetrahedron.

## 9.2 The airy and the watery vortex

We shall look at a special case of the hybrid tetrahedron where the two real points are considerably displaced from each other: one is at infinity, the other local. The two imaginary points lie on the real horizontal line at infinity. The two real basic lines in Figure 9.12 – I will now call them a and z as they provide the core of the structures – are very far apart; the line a here is vertical and local while the line z is the horizon at infinity. On a there is a lower point B, and a point A, above at infinity. The plane β *(beta)* is horizontal and goes through point B. Now imagine the point/line/plane triple moving in this space. It will move in a series of

Opposite, from top left:
Figure 9.12 An airy vortex
Figure 9.13 The watery vortex
Figures 9.14 and 9.15 show a view from
above with equiangular spirals, and a
profile showing hyperbolae fields
Figure 9.16 The watery vortex with the
spiral curve emphasised

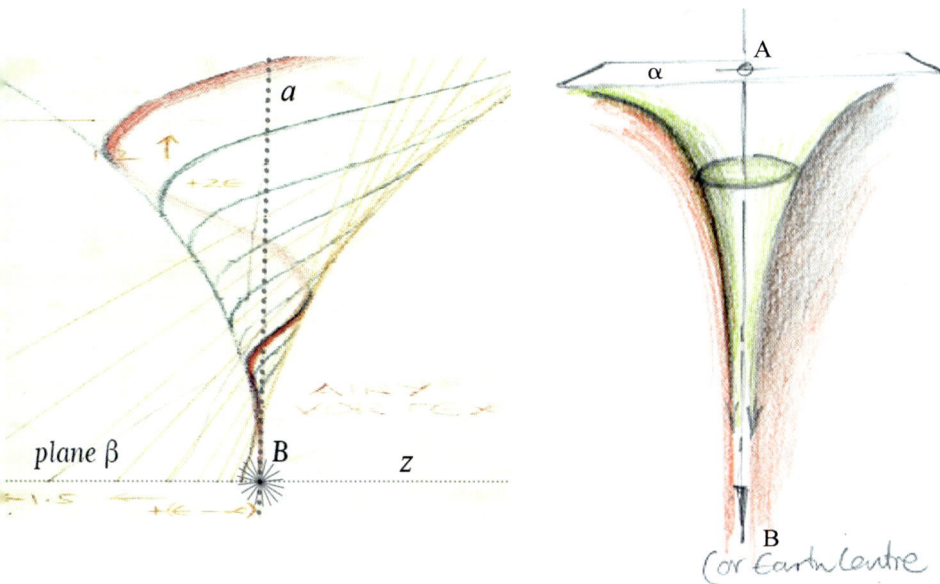

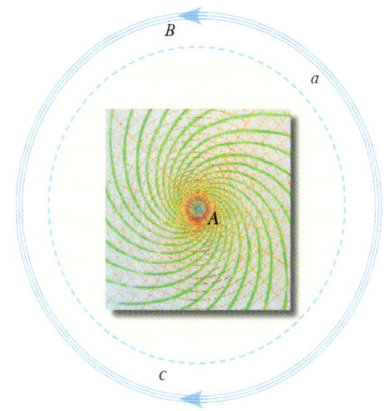

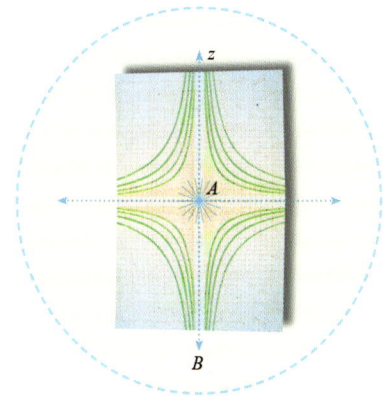

beautiful asymmetrical vortex curves, spinning about the line *a* and striving towards the line *z*. Lawrence Edwards called this form an *airy* vortex, to distinguish it from the *watery* vortex we shall describe below. In their form, they seem qualitatively to relate to these elements.

To obtain the watery vortex a number of the key elements now take completely opposite roles. In Figure 9.13 Point *B* goes off down to infinity and with it plane *β*. Point *A* now takes centre stage with plane *α (alpha)* through it. Line *a* is moved to infinity and line *z* is central to the image and perpendicular to line *a*. We need both profile and plan to see how this works. The plan is the equiangular spiral (Figure 9.14), and the profile (Figure 9.15) is a field of hyperbolae. Note again the growth measures along all axes. Put these two together interactively and a beautiful curve arises, the watery vortex. In Figure 9.16 the spiral curve is emphasised.

Lawrence Edwards said of this: 'it is the largest tetrahedron that it is possible to have: they don't come any larger. It spans all space. In one of my less prosaic moments I named it the cosmic tetrahedron.' (*The Vortex of Life*, p. 151) The drawing is only one half of the transformation; the other half is of course above the horizontal plane, and I have shown only *one* surface, while there is really an infinitude of them.

My own interest in this form stemmed mainly, at least initially, from running a experiment in our back yard to check whether an actual vortex in water is the same as the geometric watery vortex. Before 1981 I sent a few photographs to Lawrence of some water

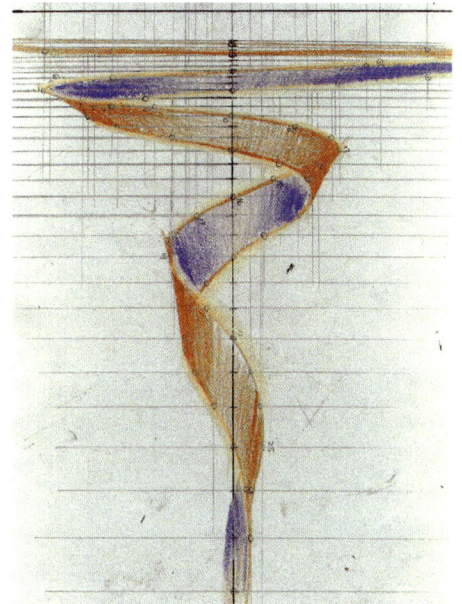

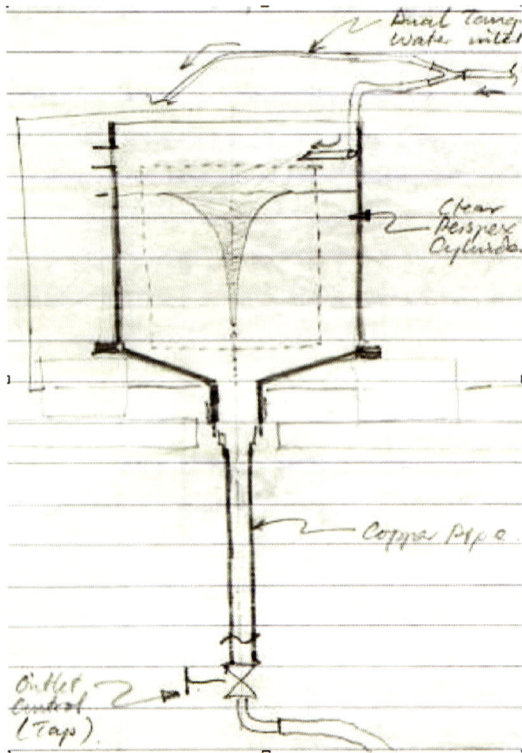

*Figure 9.17 Sketch of experimental set-up for photographing vortices*
*Figure 9.18 Vortices at varying flow rates*

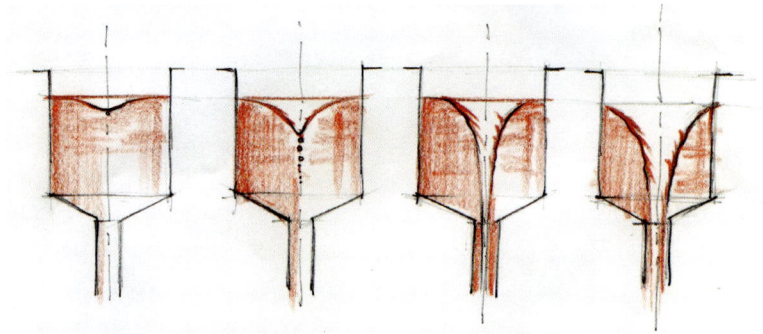

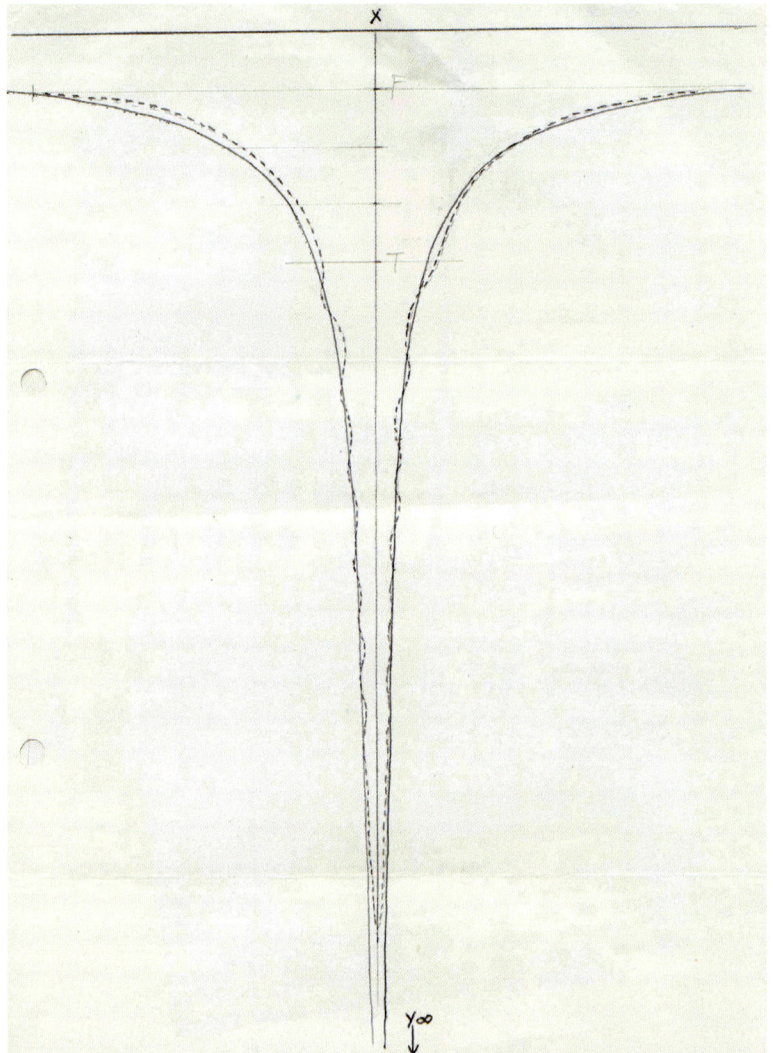

*Figure 9.19 Part of a letter from Lawrence Edwards, showing the calculated and actual profile of a vortex*

vortices in a Perspex vessel, which the company I worked with at that time had allowed me to have (Figure 9.17).

Lawrence analysed these pictures and shared his results. He found that the profiles were very close indeed to a path curve form, despite the different flow rates giving quite different looking outlines (Figure 9.18). This was encouraging. Lawrence sent me a copy of a profile he traced from such a photo and a calculated profile (Figure 9.19). The correspondence of calculated profile and traced profile was good, apart from the ripples. Unfortunately I never actually saw the photos that Lawrence used, as I sent him the original negatives. I have a couple of photos of a vortex form that were not sent, which show the apparatus (Figures 9.20 and 9.21).

To analyse the vortex we have to judge where the top horizontal surface is (not too difficult with water) and where the vertical

*Figures 9.20 and 9.21 Apparatus used*
*Figure 9.22 Inserting perpendicular axes on the photograph of a vortex [see panel overleaf]*

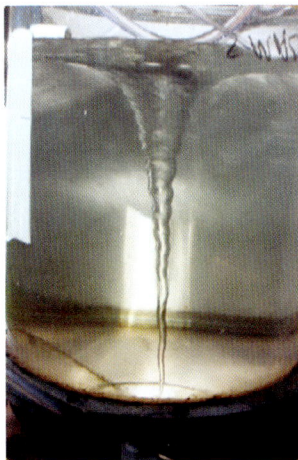

axis is. Halving the vortex width at various levels achieves a best straight line, despite the bumps in the profile. Even orienting the data photo in the first place is a guess. The intersection of these two estimated lines gives top point $X$ while point $Y$ is considered as virtually the point at infinity on the vertical axis (Figure 9.22).

I wondered whether this point $Y$ was to be thought of as at the very centre of the Earth (as a centre of gravity). Practically it would make no difference to the calculation, as the centre of the earth and infinity are both a long way off. This method of assessment assumes that the tetrahedron is 'cosmic', as Lawrence Edwards called it. This means that the measures along the vertical and horizontal axes can be taken as geometric (growth) series with a, probably different, simple multiplier. This simplifies the work significantly (see panel opposite).

Following the analysis, there seems to be a reasonable fit on the left hand side, less so on the right (Figure 9.23). For me, coupled with Edwards's results, this seemed a good indicator that these considerations of pure form were valid.

As far as the form is concerned these distances are irrelevant. The form *as such* has nothing to do with size or scale. Geometrically there is nothing that would determine any absolute sizes. There may well be other constraints from other aspects or forces that do dictate actual size, even relative size. (As an aside, I have often thought that the mean human height should be used as a measure rather than some arbitrary metre, foot or wavelength. Le Corbusier might have been getting close to some such constraints when he put forward Modular Man.)

It is a further step to tackle the spiralling. This is attempted in Chapter 13 on polar forms.

*Figure 9.23 Estimated profile of the vortex*

1 Estimate two tangents to the curve of the vortex about midway along it.

2 Mark the crossing points of these tangents on the vertical and horizontal axes.

3 Measure $M_1$ ( = 33.5 mm) and $N_1$ ( = 19.4 mm) to central axis.

4 Measure $M_2$ ( = 44.7 mm) and $N_2$ (= 84.4 mm) to top horizontal plane, $\alpha$

5 Calculate the two multipliers on the axes.
   Let $a = M_1 / N_1$ (away from point $X$), = 33.5 / 19.4,
   = 1.726
   and let $b = M_2 / N_2$ (towards point $X$) = 44.7 / 84.4,
   = 0.5296.

6 Let $\lambda_v = \ln b / \ln a$ (where $\lambda_v$ is a form factor similar to the bud/egg $\lambda$ *(lambda)* of Edwards). Hence $\lambda_v = \ln 0.5296 / \ln 1.726$,
   = −0.6355 / 0.5458, = −1.164, a negative value. This may give us an idea of a form factor for the profile.

But to give an idea of the visual correspondence we need a few more steps between $M_1$ and $N_1$ and on either side of them. How are these calculated? These were worked out using the formula for the geometric series $T_n = ar^{(n-1)}$.

If $a = 19.4$, $n = 4$, $T_4 = 33.5$, then $r = 1.199$.

From this a few more steps can be calculated. So the top series is:
   13.4, 16.2, 19.4, 23.3, 27.9, 33.5, 40.23, 48.3
   and calculating similarly for the vertical with $r = 1.237$ the vertical series is:
   29.2, 36.1, 44.7, 55.4, 68.7, 84.4, 104.4, 129.1

The next step is to join corresponding points, thus obtaining the possible tangents to the form (shown dotted in red in Figure 9.23).

Figure 9.24 A neolithic carved stone ball from Scotland, with four 'faces' suggesting a tetrahedron [Critchlow, Time Stands Still]. No one knows what these objects were used for or exactly how old they are

Figure 9.25 Plato's tetrahedron as the essence of fire

## 9.3 The real and regular tetrahedron

As we saw at the beginning of the chapter, the simplest, most fundamental form in space is the tetrahedron with four points (vertices), six lines (edges) and four planes (surfaces). This is the central form of the five regular Platonic solids and is also one of the most common forms found among the Scottish carved solid stone balls from neolithic times. Plato's *Timaeus* describes the tetrahedron as the 'being and essence of fire'. Plato perhaps saw something of the primal spatial form, the tetrahedron, and the first manifestation of creation in a primal heat.

A regular tetrahedron is one where all the surfaces are equally sized equilateral triangles and all the lines are of equal length. There are, of course, many ways of constructing such a regular tetrahedron but to take one approach, we can imagine beginning with the lines. Let us begin with two skew lines (that is, not touching), and set them perpendicular to each other. Now take another two pairs of lines, each pair perpendicular to each other. We now have *six* lines in total. We can bring them together, the three pairs all about the same centre point, or to be precise the three lines joining each skewed pair (or bridge lines, we could call them) are brought together in one point. Figure 9.26 shows, this, though it is always difficult to draw lines of three dimensions onto a flat image – there is no shape for us to recognise that creates the three-dimensional picture in our mind's eye. Next arrange the bridge lines to be mutually at right angles to each other. Now the tetrahedron appears, marked out by the six lines. In Figure 9.27 I have added in the four points as well as a hint of the

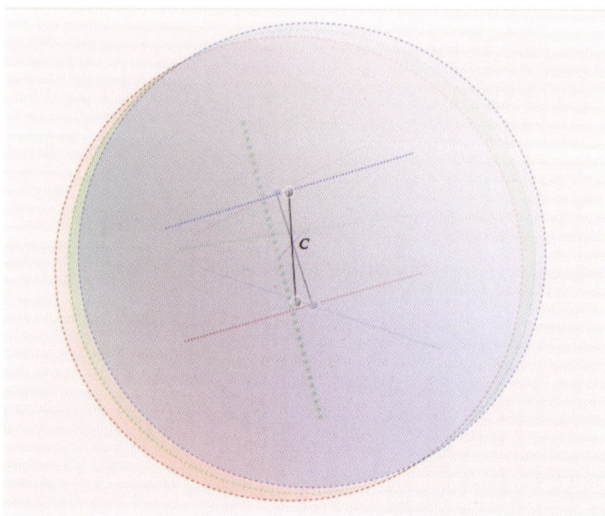

Figure 9.26 Centres of the bridging lines between each pair of perpendicular lines coinciding in a point;

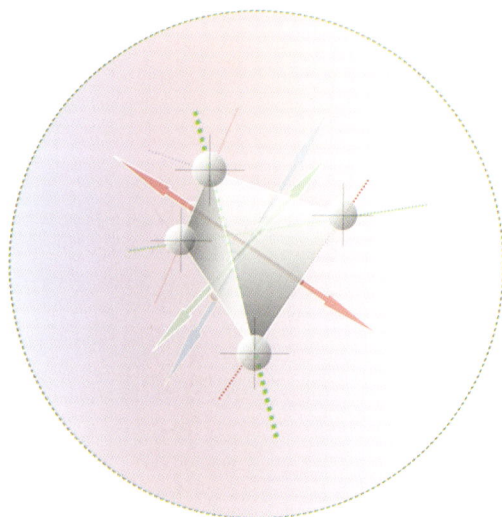

Figure 9.27 The regular, equilateral tetrahedron

planes of this tetrahedron. The whole thing is fully symmetrical in *every* direction.

In section 8.2 we used the framework of a tetrahedron to generate the forms of the vortex. In that case the tetrahedron was semi-imaginary, that is two points were real, and two points were these strange, circling 'imaginary' points, the geometric equivalent of the square root of a negative number. We shall now turn to a very special case of the tetrahedron arrangements.

## 9.4 The ultimate degenerate tetrahedron

The ultimate degeneration of the tetrahedron is when all six lines, all four planes and all four points come together. Surprisingly this can still generate forms with real life representatives. It leads to the point/line/plane triples creating a series of elliptical curves. The model shows a hint of how this might appear (Figure 9.28). The form is akin to bivalve seashells.

The geometrical forms have a common line about which the ellipse 'hinges'. When I heard of this so-called degenerate from Lawrence Edwards, it blew me away. The bivalves exhibited some natural forms – often extraordinarily subtle in their curves – that allowed them to open and close with a hinge-like mechanism, and here was the geometry demanding a straight line to give such forms! This form was introduced to me in the plane where it is not the tetrahedron that degenerates but a triangle (Figure 9.30). This is elaborated in the next chapter.

Is there a hint of morphological *evolution* here? The bi-valves are an early creature in the fossil records of the earth. Geometrically, only a very simple, special degenerate manifestation of the general tetrahedron appears first or very early. Other potential forms are not yet visible but are inherent and concealed.

In conclusion we can say that a sequence of structuring occurs within and throughout the tetrahedron. *Rhythms* arise when the line transforms into itself. *Shapes* arise in two dimensions when the plane (or point) transforms into itself. And three-dimensional *forms* arise when space does its special thing; that is, when it too transforms into itself. We have then:

*rhythms - shapes - forms*

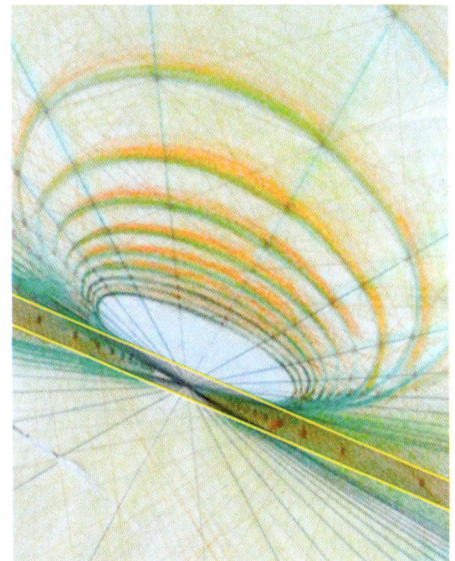

*Figure 9.28 Model of curves generated by the ultimate degenerate tetrahedron*
*Figure 9.29 Simple bi-valve shell*
*Figure 9.30 Form arising from a degenerate triangle*

# 10. Convex Path Curves

## 10.1 A general, real triangle

We saw in section 8.3 that the forms associated with point/line transformations can be called path curves. These curves are the inevitable paths of a point/line pair moving in a field based on some kind of triangle. Perhaps they should be called *envelope* curves (of tangent lines) as well as path curves (of points).

Within the all-real triangle, in other words the everyday one we are used to, we can find how the plane transforms into itself. There are always three invariant points and three invariant lines, that is points and lines that themselves do not change. But the rest of the plane does change. It can be thought of as in constant motion with tangent points and lines following precise curves that cover the entire plane. Something akin to it can be found in the movement of water. We see it, for instance, in the form of an eddy, the form of a standing wave in a river, or in a waterfall where the form remains while the matter (the water) rushes through it.

Our first example is for a general scalene triangle (sides all different lengths). The three invariant (fixed) points and lines overlay the graphic (Figure 10.1).

*Figure 10.1 Triangle path curves*

## Constructing a path curve

1  Select three invariant points, *ABC,* and connect
   them with lines *a, b* and *c* (Figure 10.2).

2  Select any two points, *M* and *M'*, somewhere
   within the outline *ABC.* They could be anywhere,
   but it is convenient if they are within the triangle.
   Imagine that *M* moves (translates) to *M'*. This
   creates line *m* (Figure 10.3). By this last choice
   the whole diagram is now determined: it is simply
   not yet visible.

3  Join *AM* and extend to meet *P* on line *a.* Join
     *AM'* and extend to meet *Q* on line *a.* Join *CM* and
   *CM'* and extend to meet *P'* and *Q'*
   respectively on *c.* So the result is that line *AM*
   swings to *AM'* and line *CM* swings to *CM'*
   simultaneously (Figure 10.4).

4  Establish a growth measure (see section 7.2) of
   points along *BC* using *P* and *Q* as the two initial
   points as in Figure 10.5. Similarly establish a
   growth measure along *AB* using *P'* and *Q'* as
   the two initial points.

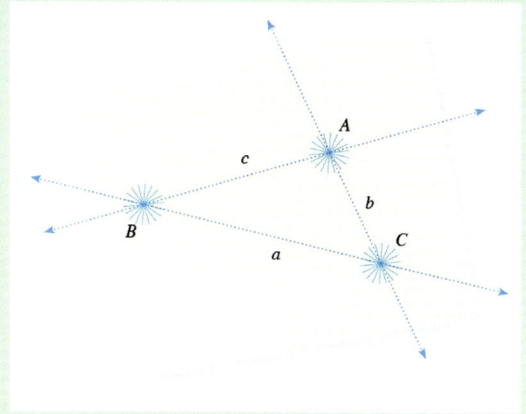

Figure 10.2

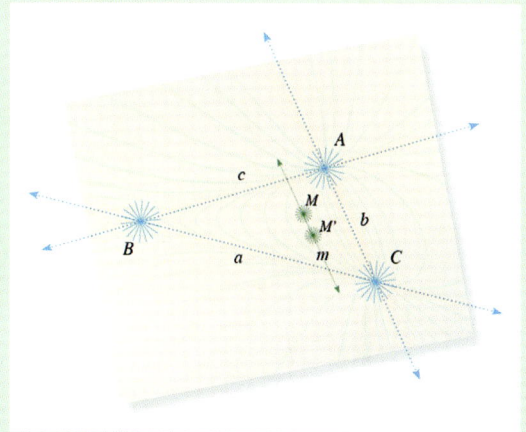

Figure 10.3

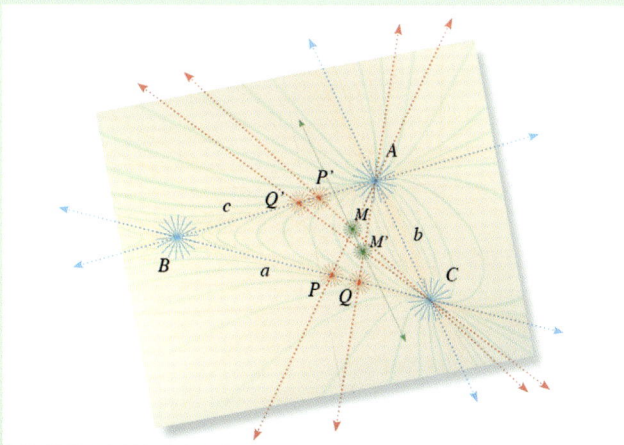

Figure 10.4

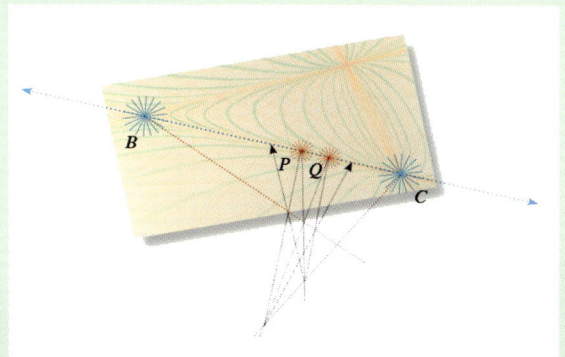

Figure 10.5

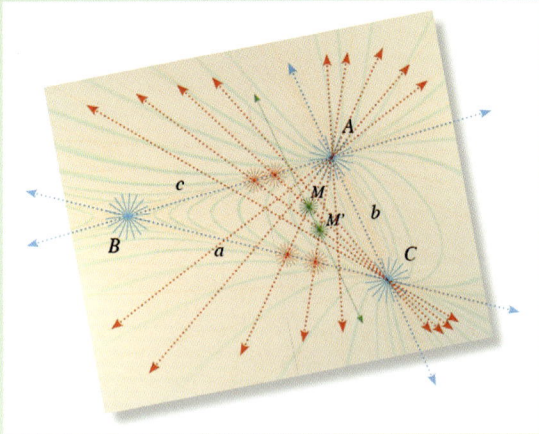

Figure 10.6

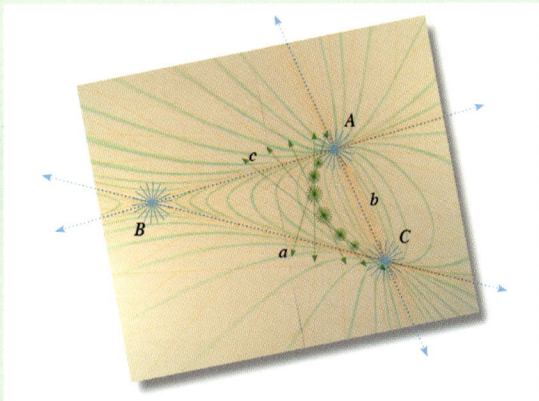

Figure 10.7

5 Draw in the family of rays from *A* and from *C* to their respective measures. Only a few more are shown in Figure 10.6.

6 Continue the path that *M* follows, both towards *C* and to *A*. The curve that these points make is a *path curve*. A few such points as well as the associated tangent lines are indicated in Figure 10.7.

7 It is possible to continue the profiles beyond *A* and *C*, and to draw many of the other curves that make up this field. All we have to do is to join across all the little quadrilaterals (intersections of all the orange lines) in the same direction. This results in the family of curves we saw at the beginning of the chapter (Figure 10.1).

## 10.2 A real triangle with one point at infinity

What about other triangles? One way of making the general triangle into a special one is simply moving point *B* off to infinity so lines *c* and *a* become parallel, and make line *b* perpendicular to *c* and *a* (Figure 10.8). Despite its appearance this is still a triangle.

The curve construction is the same in method as for the previous triangle (Figures 10.3 to 10.7). One difference is that there has been an apparent simplification. The measures along the lines *a* and *c* are now a geometric series, a special case of growth measure where there is a constant multiplier (Figure 10.9). Here the curves have become bilaterally symmetrical about the line *b*. This special case has to do with both distance (*B* at infinity) and angle (line *AC* perpendicular to the two parallel lines. This is an extremely important case and will be developed in some of its implications for natural forms later in this chapter.

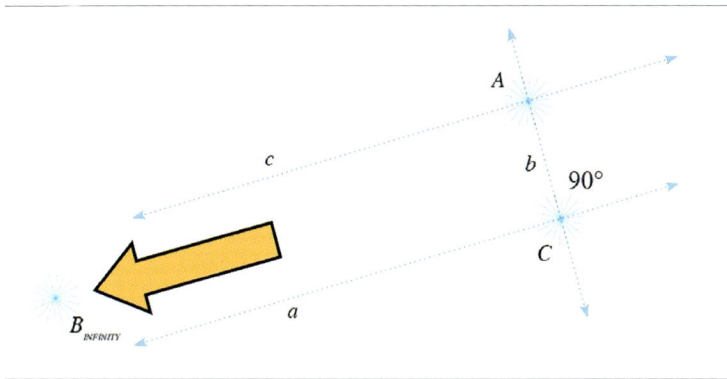

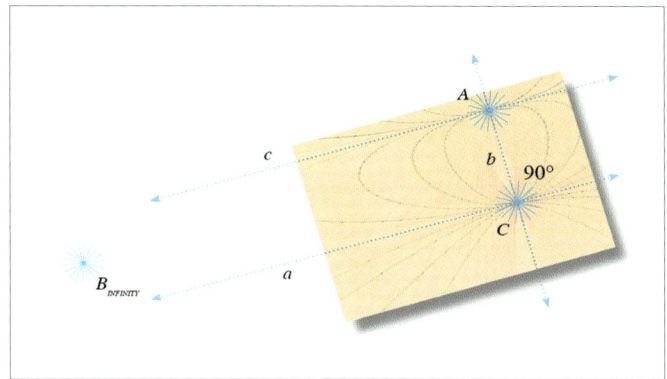

This is a peculiarly interesting triangle as all the elements have melted into what looks simply like a point/line pair: the points *A, B* and *C* have shrunk into one, now called point *A,* and the lines *a, b* and *c* have merged into a single line, called *a* (Figure 10.10 Triple point and line). How can this possibly have shapes associated with it? Lawrence Edwards shows in *The Vortex of Life* (p. 37) how the measure in the line is a step measure of points, and in the point is a *step measure* of lines. Step measure is a special case of growth measure. We simply set up these two step measures in point *A* and on line *a,* and throw in a point/line pair, and then track the inevitable path (Figure 10.11). The curves are a family of ellipses all touching point *A* and tangent to line *a*. No one ellipse is the same as another in size or orientation, yet they all belong to the same infinitely extensive field of form.

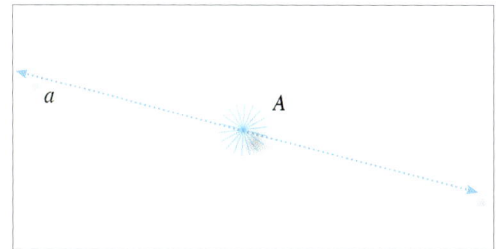

*Figure 10.8 A triangle with one point at infinity, and lines a and c parallel*
*Figure 10.9 Symmetrical path curves*
*Figure 10.10 A triangle with all points and lines coincident*

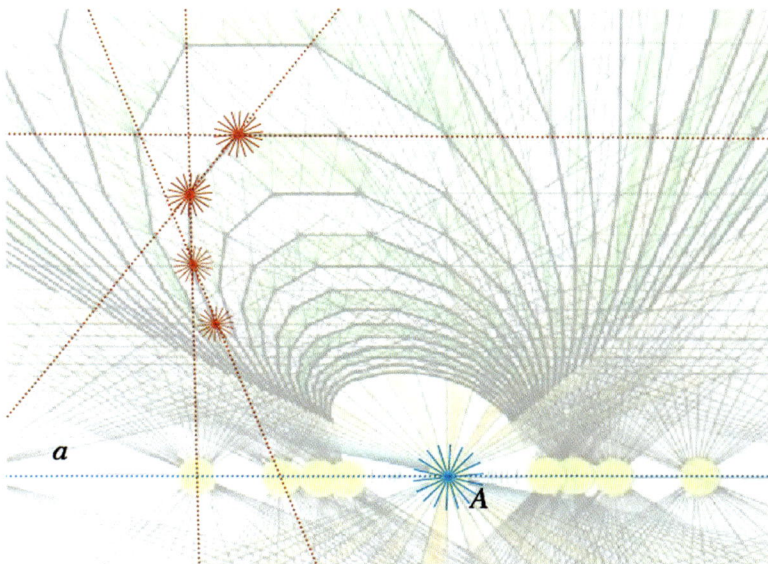

*Figure 10.11 A family of ellipses arising from step measure*

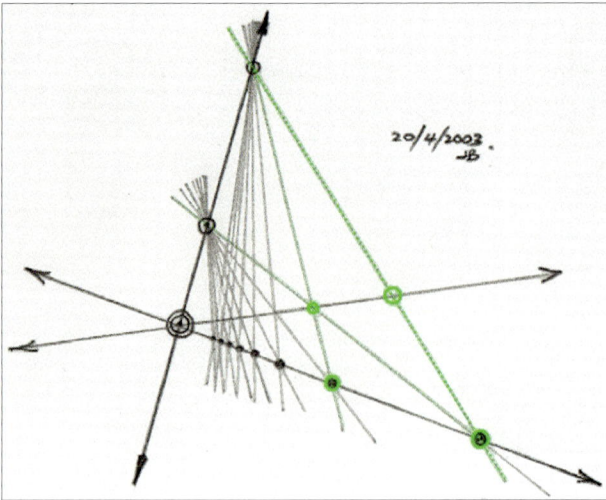

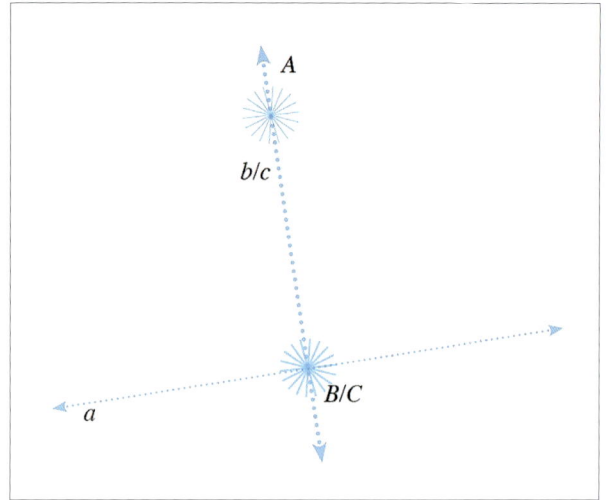

Figure 10.12 Step measure in two points
and two lines
Figure 10.13 A real triangle with two
points and two lines coincident

This is another profile case of interest, a sort of midway case. This is where point *A* remains separate, and *B* and *C* merge. This means, of course, that lines *b* and *c* merge while *a* remains separate (Figure 10.13). Line *a* does not have to be perpendicular to line *b/c* but I have made it so. We have step measures in line *a* and growth measure along line *b/c*. I just give a single example here. (This is all carefully detailed in Edwards, *Projective Geometry*, pp. 227ff.) A growth measure is needed along line *b/c* focused on the points *A* and *B/C*, and a step measure along *a*, and focussed on *B/C* (Figure 10.14).

Figure 10.14 Middle of the road paths
Figure 10.15 A real triangle with two
points at infinity and two perpendicular
lines creates vortex profiles

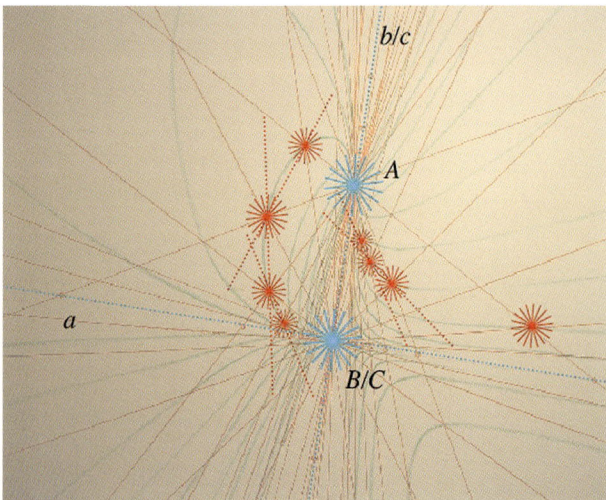

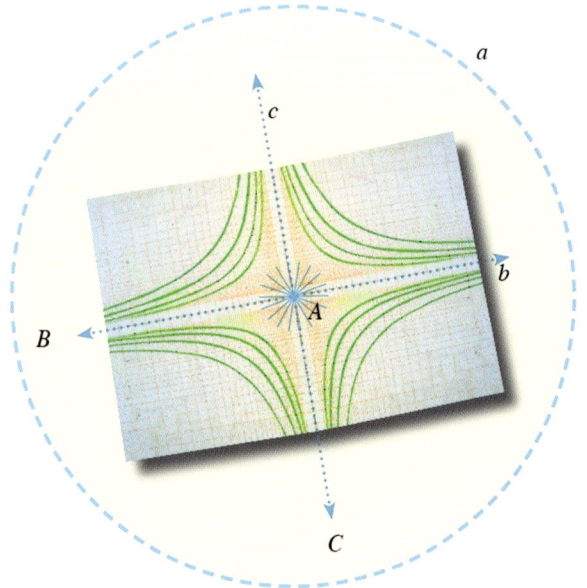

Another important case is where two points, *B* and *C*, move off to infinity and *a* becomes the line at infinity. We put point *A* centre stage with mutually perpendicular lines *b* and *c* through it. The path curves that are the consequence of the transformation inherent in this arrangement are like a vortex profile (Figure 10.15). These curves are a particular kind of vortex profile; they are simple hyperbolae.

If the curves are drawn across the little rectangles on the other diagonal we get the curves Edwards calls an airy vortex (Figure 10.16).

## 10.3 Semi-imaginary or complex triangles

We saw previously (in section 9.2) that imaginary numbers (the square roots of negative numbers) come in pairs, and in geometry are expressed as being in circular motion. As imaginaries always come in pairs, it is impossible to have all three points or lines of a triangle being imaginary, one point and line always remains real. So they are really just semi-imaginary, or complex (containing both real and imaginary elements).

Let us take a triangle and let points *B* and *C*, and lines *b* and *c*, rotate in circling measure about the real point *A*. And line *a* remains fixed, stuck, still and static. Points *B* and *C*, can be *imagined* to flow, to translate, along line *a*. Lines *b* and *c* (not shown specifically) are imagined to rotate in point *A*. They rotate in *circling* measure and the pair of points translates in circling measure of points as well.

The resulting forms are the spirals we came across earlier in Chapter 8. Elaborating Figure 10.13, we now see these spirals in a wider context (Figure 10.17). Note the points *B* and *C* are in flow and counterflow.

The next triangle is even more radical. The triangle is as before, but we now move the real line to infinity. We have come across the spiral curves this produces in Chapter 8 (Figure 8.19). We note that *B* and *C* are on *a*, the line at infinity, but are in an even circling measure, that is in equal steps, moving in opposite directions. Along the lines radiating from centre point *A*, are geometric measures (the special case of growth measure that happens when either *B* or *C* have moved to infinity). Hence the ever-expanding concentric circles about point *A*, with radii increasing in geometric series. The curves arising from this arrangement are equiangular spirals (Figure 10.18).

There are of course *two* sets of these, one clockwise and the other anticlockwise. These are overlaid over each other in Figure 10.19. This begins to look like a sunflower, though in nature the number of spirals clockwise is not usually the same as those anticlockwise.

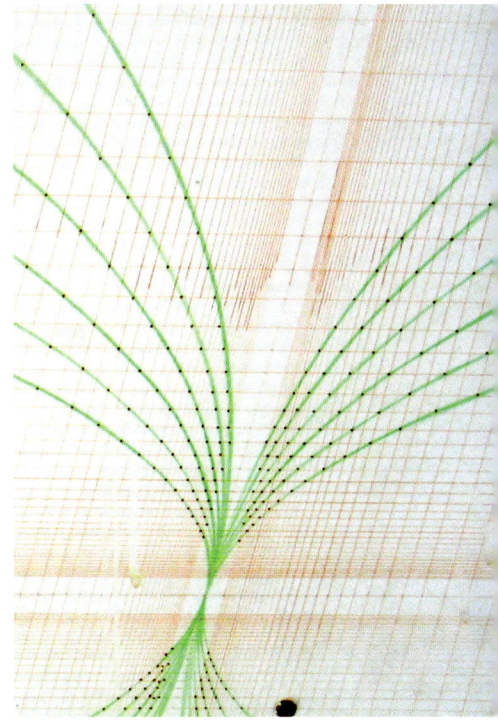

*Figure 10.16 Airy vortex*

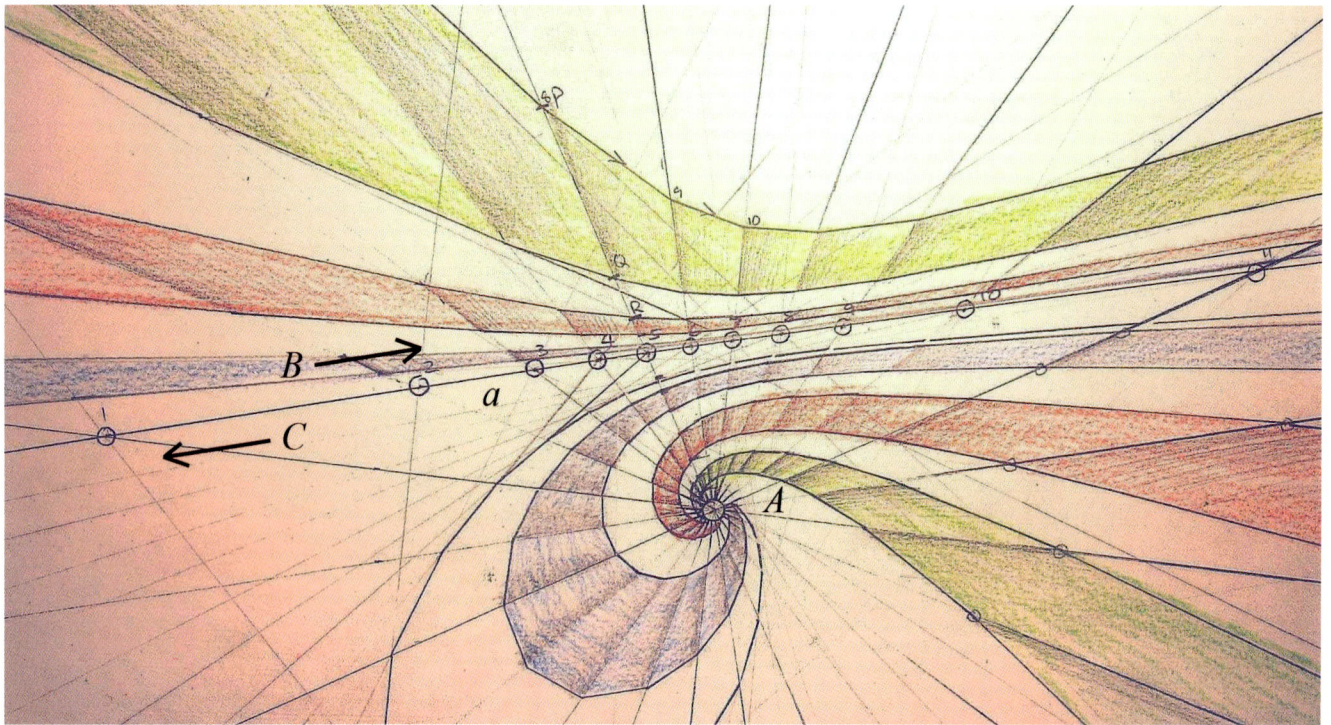

Figure 10.17 The field resulting from an imaginary triangle with one real point and one real line

If the spirals are very slow they tend more and more to *circles*. So even concentric circles are path curves. If, on the other hand, they get steeper and steeper, they tend toward the *straight* radii through point A. They effectively range from straight radiating lines to concentric circles through every possible equiangular spiral in between (Figure 10.20).

Figure 10.18 Equiangular spirals arising from a complex triangle with one real point (centre), one real line (at infinity) and two points and lines circling
Figure 10.19 The double spirals
Figure 10.20 The range of spirals from radial lines to circle

## 10.4 The bud

We shall now look at the two principal shapes that are generated within a three-dimensional, hybrid tetrahedron, that is part real, part imaginary. One of these shapes, the general vortex, was briefly explored earlier. The other is the convex form that includes the shape of eggs or buds. This form is far from limited to eggs and leaf and flower buds, also appearing in tree profiles and possibly sea urchins.

To begin with, it is simpler to look at profiles in the plane than the spatial forms. In the section 10.2 we looked at the field produced by triangle with one point at infinity, and lines *a* and *c* parallel. It is shown again here (Figure 10.21).

In the mid 1970s I had the opportunity to work with Lawrence Edwards while he was visiting Australia for some months. He had a number of photographs of some small plant buds – I cannot recall what species the buds were, and it doesn't really matter. It was the form *as such* we were examining. The first step was to lay tracing paper over the photographs and carefully trace the profiles. The outlines were traced on one side of the paper and then it was turned over to do the actual measuring. This enabled the original outline to remain intact if ever review was needed in the measuring.

To measure it, the endpoints have to be marked. The bud top is usually clear, but the lower pole has to be an estimate due to the uncertain merging with the connecting stalk. Then an estimated central axis is drawn between these poles. Many buds are virtually vertical, so this is not too difficult. The height is divided into eight and the measurements are made at these divisions. The details are shown in the panel below.

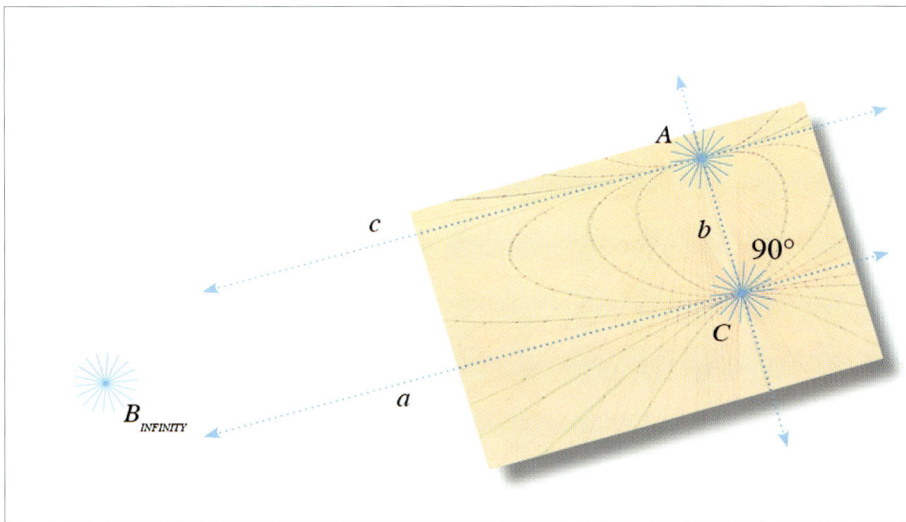

*Figure 10.21 The field produced by triangle with one point at infinity*

Figure 10.22 Photograph of some small plant buds

Figure 10.23 An early tracing and measurements from a photograph

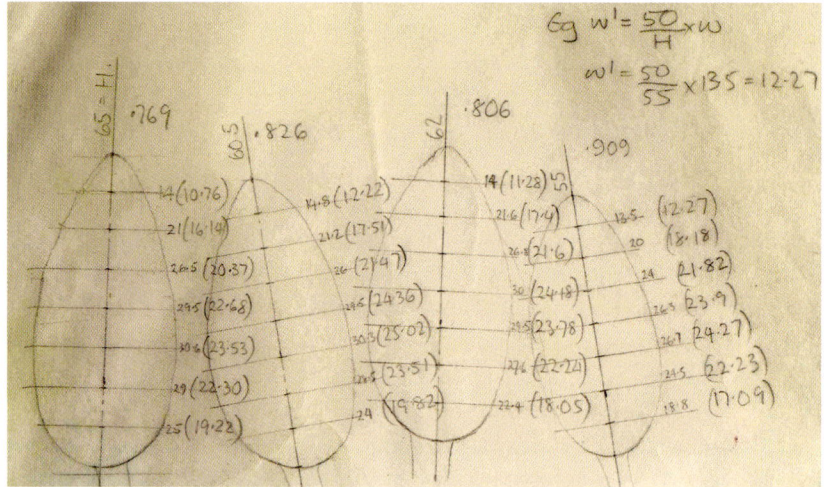

Figure 10.24 Seven levels

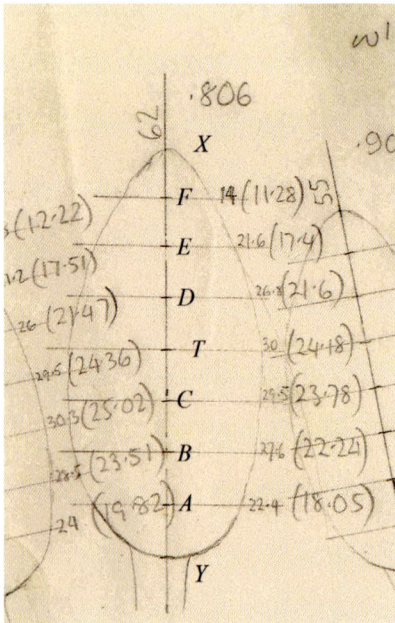

## Analysis

1 Estimate a top pole, point $X$, and a bottom pole, $Y$ (Figure 10.24).

2 Then draw an estimated central axis between these poles, $X$ and $Y$.

3 Then divide this height, $XY$, ($XY = 62$ mm) into eight spaces, labelled (from bottom) $A, B, C, T, D, E$ and $F$. Then draw these seven levels $A, B, C, T, D, E$ and $F$ at right angles to the central line, $XY$.

4 Now measure the diameters at each level. For instance the bud diameter at $T$ was measured as 30 mm.

5 Now all these diameters were scaled to a standard height of 100 mm. A typical calculation shown in Figure 10.24 would be:
Scaled diameter = $(100 / H) \times D'$
then scaled radius is half the diameter, so
Scaled radius = $(50 / H) \times D'$
Assume measured $D' = T = 30$, and $H = 62$ mm, and standard height is 100 mm
then scaled radius = $(50 / 62) \times 30$
so that at level $T$ Scaled radius = 24.2 mm.

6 This is done for all of the seven levels of this one bud.

*Finding λ (lambda) and plotting the ideal profile*

1 These scaled radii (Figure 10.24) were:
   A = 18.05
   B = 22.24
   C = 23.78
   T = 24.18
   D = 21.60
   E = 17.40
   F = 11.28

2 This is repeated for similar buds and the average scaled radii values are listed.

3 λ is the ratio of the logarithms of the ratio of the mean radius at each level to the radius at the central level *(T)*

4 Then the average λ-value of the seven levels is calculated. In the measured case shown here the average is:
   λ = (1.319 + 1.335 + 1.399 + 1.874 + 1.972 + 1.998) / 6
   Hence λ = 1.649

5 To plot such a curve, a geometric series is needed along the top horizontal line with multiplier *a,* and along the bottom horizontal line with multiplier *b* (Figure 10.25) such that:
   λ = log a / log b
   If we let *a* = 1.2 (this can be freely chosen) and λ = 1.649 then *b* is totally determined, and will be found to be 1.117.

6 The curve we are looking for runs through the point where *T* = 23.738 (the average of a number of buds). To find starting point on top and bottom lines we simply double the value for *T* (2 × 23.738 = 47.476). Now calculate and mark a geometric series along the top line where *a* = 1.2 (chosen) and along the bottom where *b* = 1.117 (determined above).

7 The path curve is the progressive intersection of the lines rotating in the same direction in *X* and in *Y* (Figure 10.27). This is the transformation of this plane into itself based on this particular triangle with one point at infinity and with two lines perpendicular to the third, vertical line.

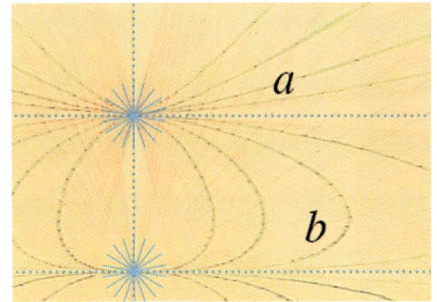

*Figure 10.25 Multipliers a and b*

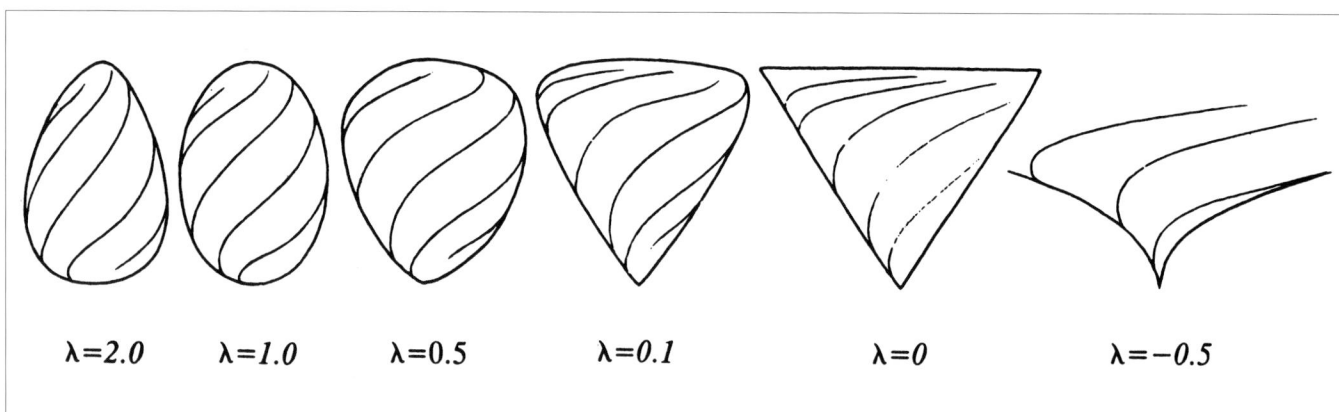

λ=2.0        λ=1.0        λ=0.5        λ=0.1        λ=0        λ=−0.5

*Figure 10.26 Spectrum of forms expressed in λ-values (from Edwards,* The Vortex of Life)

How can we check whether this form coincides with the path curve? In total there were six buds. The same procedure was applied to all the six, and the results were then averaged. Details are in the panel.

Edwards developed a method to calculate the form factor, which he called λ *(lambda)*. The method is in Appendix 3 of *The Vortex of Life*, 2006 edition and will not be elaborated here. However, the form factor can still be described pictorially. It is characterised by a degree of bluntness or sharpness. The profiles of these little buds are sharper at the top and more rounded at the base. Note also that the sharper the top, the blunter the base will be. The two are related, and you cannot have an egg-form (or bud-form) that is very sharp at one end and also tending to sharpness at the other. Edwards showed this form spectrum (Figure 10.26) with a range λ-values from +2 to −0.5.

To plot such a curve a geometric series is needed along the top horizontal line, and another geometric series along the bottom horizontal line. These two series are related to each other through λ, the form factor. A pencil of lines is drawn through the points of the series on the line opposite *X* or *Y,* and this gives a grid to plot the curve (Figure 10.27).

The data points coincide reasonably well with the path curve points. It was that correlation that got me seriously interested in this work, as it was *idea* working into *phenomena.* Coming from an engineering background, I felt here was something very important. Here was the living organic corresponding to a geometric construct derived from active movement. The profile of the living and growing bud seemed to be part of a dynamic field, a *field of form.* This in fact was the title of Edwards's book in 1982, which he later revised and published as *The Vortex of Life.*

These fields may be akin to what Rupert Sheldrake describes as morphic fields. These fields concentrate around the bud,

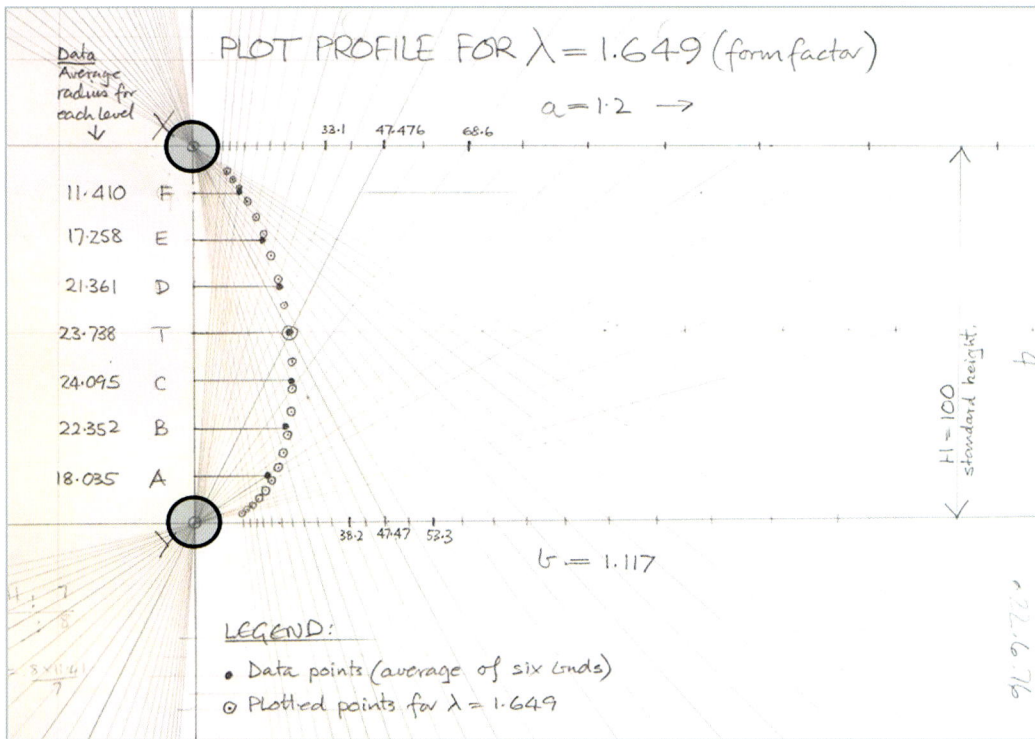

Figure 10.27 Plotted profile of a bud shape

but extend over the entire plane, or rather through the whole of space. The physical artefact, the bud in this case, expresses the merest fragment of the field, yet seems to be determined by it and is certainly embedded in it. This field not only holds the one form, but has related forms, none the same, yet all related in a whole (Figure 10.28). And between each drawn curve there is an infinitude of others.

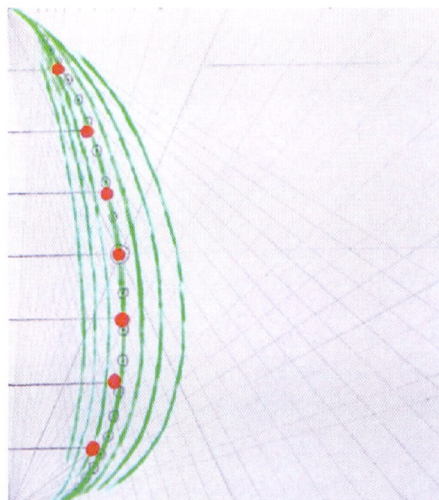

Figure 10.28 A few of the path curves from the same field

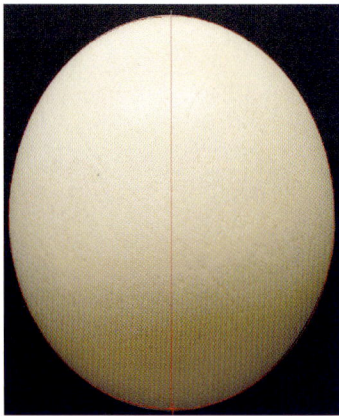

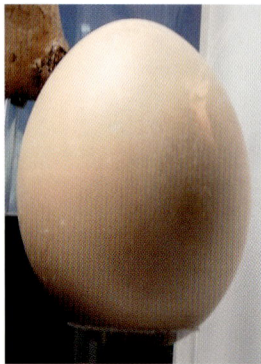

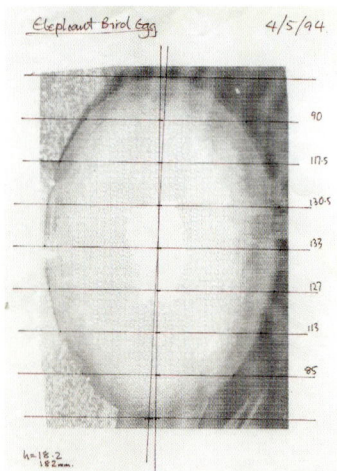

## 10.5 Eggs

How widespread are these forms? Are there other organic structures that followed the form of path curves? Is our world nothing but path-curves? I began to explore.

The egg form has a perfection and harmony about it, even a beauty. It is a pure, continuous convex form in space. And there are legions of them in all sizes and colours. A picture and a profile of an ostrich egg came my way recently. I was particularly interested in this egg as it had a hint of spiralling on it too (more of that in Chapter 14). The profile was an exceptionally good fit (Figure 10.29, thin red curve). The $\lambda$-value is very close to 1, which is an ellipse.

I came across a giant egg in the Western Australian Museum. It is believed to be the egg of the elephant bird that became extinct thousands of years ago in Madagascar. There are stories about how two of these came to the west coast of Australia. One was found quite recently by a boy who kept it hidden for a while fearing it would be taken from him in the 'cause of science'. Negotiations eventually saw it in the museum. It is large, about 30 cm (1 ft) long. It is the largest birds egg I have ever come across; even dinosaur eggs are smaller than this. However, the egg is a good path curve, too. Analysis showed the mean $\lambda$-value to be 1.072, and a standard deviation of 1.8%. As the $\lambda$-value is so close to 1, the shape is almost an ellipse.

Note that the $\lambda$-value was greater than 1. If the analysis had been carried out with the form the other way round, the $\lambda$-value would have been less than one (in fact 0.933, the reciprocal

*Figure 10.29 Ostrich egg showing a very good fit* [Nick Thomas/John Wilkes]
*Figure 10.30 Elephant bird's egg in Melbourne Museum*
*Figure 10.31 The data on a poor photocopy of the elephant bird egg*

*Figure 10.32 Emu*

of 1.072). This was worked on before I had decided on the convention that all animal forms should have the sharper end down, resulting in a λ-value between 0 and 1, and those of the plant world would have the sharper end up (as in most buds), resulting in a λ-value between 1 and infinity.

Another large bird's egg is that of the emu. The emu, a native of Australia, is flightless, can stand as high as a human, and can be an unfriendly creature. The egg is 13 cm long, and a dark greenish colour (it is often decorated for tourists by grinding away the dark outer layer leaving white beneath). The egg turns out to be an excellent path curve profile with a weighted mean λ-value of 0.932. The weighted mean weights the more sensitive dimensions at the ends. The mean λ deviation was 3.6% and the mean radius deviation was 1.43%.

In Australia we have two rather special mammals that lay eggs. One is the platypus and the other the echidna. These are known as monotremes, one poisonous and the other prickly.

When I met Mervyn Griffiths in Canberra he had already done a lot of work on the echidna, published whole erudite books on them, and worked for years in a government research establishment in Australia. He happened to have a photograph of an egg as it emerged from the little beast, and gave me a copy of this. I was lucky as these creatures are strongly protected and I did not think I had much chance of getting an image of such an egg without some professional help.

The analysis showed a λ-value of 0.846, and a mean radius deviation of 1.83%. Not bad for a mere echidna.

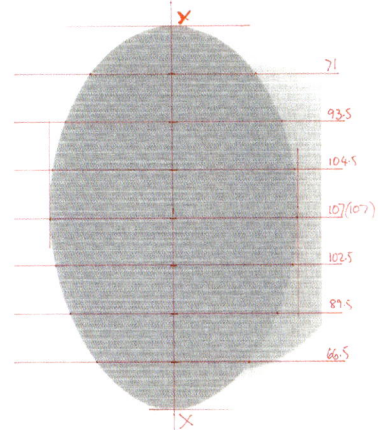

emu, 20/3/1994

| level | dia(mm) D | normal/a D/T | lambda L | weight L W*L | lambda de abs dev | ideal rad Dt | DI*T | height H | radius dev abs dev | % dev |
|---|---|---|---|---|---|---|---|---|---|---|
| Y | | | | | | | 0.0 | 158.0 | | |
| F | 71.00 | 0.6636 | 0.993 | 3.974 | 0.062 | 0.685 | 73.2 | 138.2 | 2.249 | 3.0704 |
| E | 93.50 | 0.8736 | 0.968 | 1.836 | 0.036 | 0.883 | 94.5 | 115.5 | 0.981 | 1.0380 |
| D | 104.50 | 0.9766 | 0.935 | 0.935 | 0.003 | 0.977 | 104.5 | 98.0 | 0.041 | 0.0398 |
| T | 107.00 | 1.0000 | | | | 1.000 | 107.0 | 79.0 | | |
| C | 102.50 | 0.9579 | 0.920 | 0.920 | 0.012 | 0.960 | 102.7 | 59.2 | 0.172 | 0.1671 |
| B | 89.50 | 0.8364 | 0.881 | 1.762 | 0.051 | 0.849 | 90.9 | 39.5 | 1.384 | 1.5225 |
| A | 96.50 | 0.6215 | 0.880 | 3.519 | 0.052 | 0.639 | 68.4 | 19.6 | 1.682 | 2.7526 |
| X | | | | | | | 0.0 | 0.0 | | |
| means | | 0.929 | 0.932 | 3.6 | | | | | | 1.432 |

l= .932
h=100
a= .99
b= .990677
n1= 50
n2= 0

*Figure 10.33 Emu egg*
*Figure 10.34 Measurements of the emu egg*
*Figure 10.35 Calculations*

*Figure 10.36 An echidna: Australian egg-laying mammal [Fir0002/Flagstaffotos, reproduced under GFDL licence]*

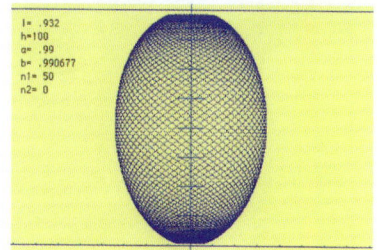

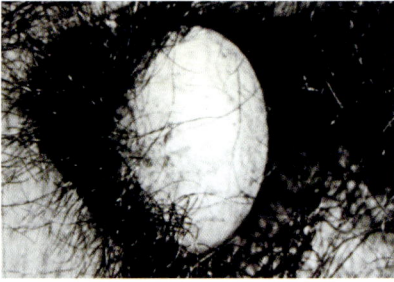

*Figure 10.37 Echidna egg*

A picture of the platypus egg was hard to find. Eventually I did find a picture, complete with scale, in a marvellous article on the platypus by Jack Green in an issue of *Australian Geographic* magazine. The egg was almost spherical, and about 15 mm in length. The picture was enlarged, measured and analysed, and the egg was found to a $\lambda$-value of 0.979. The theoretical and the actual were then compared by visual overlay, and the two were barely distinguishable.

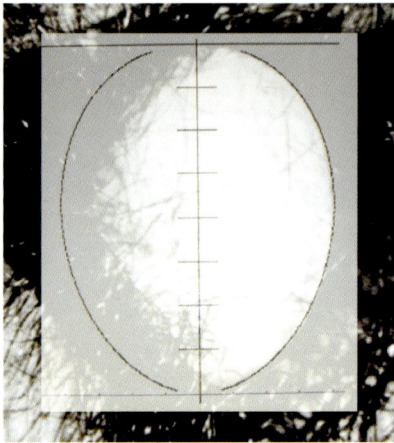

*Figure 10.38 A good fit of the Echidna egg*

*Figure 10.39 A platypus swimming in Broken River, Queensland [Wikipedia]*

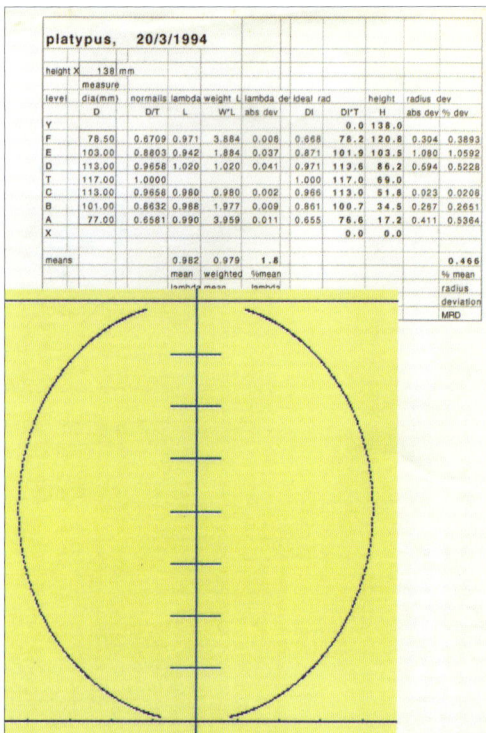

**platypus, 20/3/1994**

height X   138 mm

| level | measure dia(mm) D | normals D/T | lambda L | weight W*L | lambda dev abs dev | ideal DI | DI*T | H height | radius dev abs dev | % dev |
|---|---|---|---|---|---|---|---|---|---|---|
| Y | | | | | | | 0.0 | 138.0 | | |
| F | 78.50 | 0.6709 | 0.971 | 3.884 | 0.006 | 0.668 | 78.2 | 120.8 | 0.304 | 0.3893 |
| E | 103.00 | 0.8803 | 0.942 | 1.884 | 0.037 | 0.871 | 101.9 | 103.5 | 1.080 | 1.0592 |
| D | 113.00 | 0.9658 | 1.020 | 1.020 | 0.041 | 0.971 | 113.6 | 86.2 | 0.594 | 0.5228 |
| T | 117.00 | 1.0000 | | | | 1.000 | 117.0 | 69.0 | | |
| C | 113.00 | 0.9658 | 0.980 | 0.980 | 0.002 | 0.966 | 113.0 | 51.8 | 0.923 | 0.0208 |
| B | 101.00 | 0.8632 | 0.988 | 1.977 | 0.009 | 0.861 | 100.7 | 34.5 | 0.267 | 0.2651 |
| A | 77.00 | 0.6581 | 0.990 | 3.959 | 0.011 | 0.655 | 76.6 | 17.2 | 0.411 | 0.5364 |
| X | | | | | | | 0.0 | 0.0 | | |
| | | | | | | | | | | |
| means | | 0.982 mean lambda | 0.979 weighted mean lambda | 1.8 %mean | | | | | | 0.466 % mean radius deviation MRD |

*Figure 10.40 Platypus egg calculations*

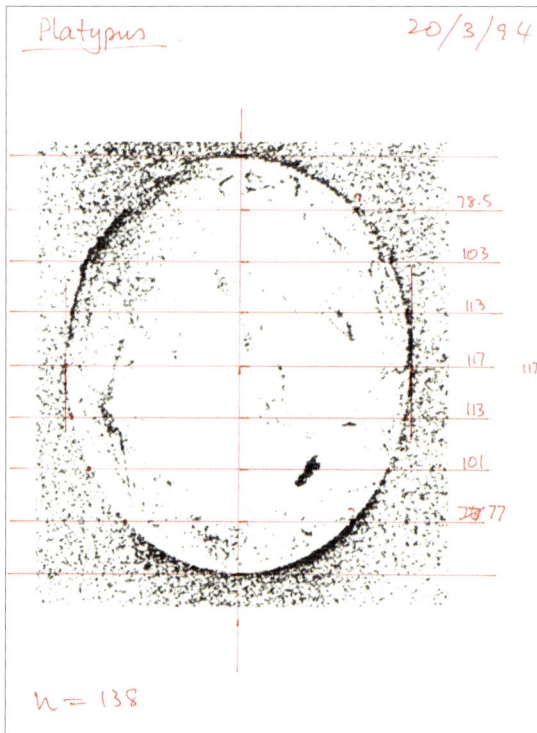

Platypus                          20/3/94

78.5
103
113
117        117
113
101
77 77

h = 138

*Figure 10.41 Platypus egg overlay*

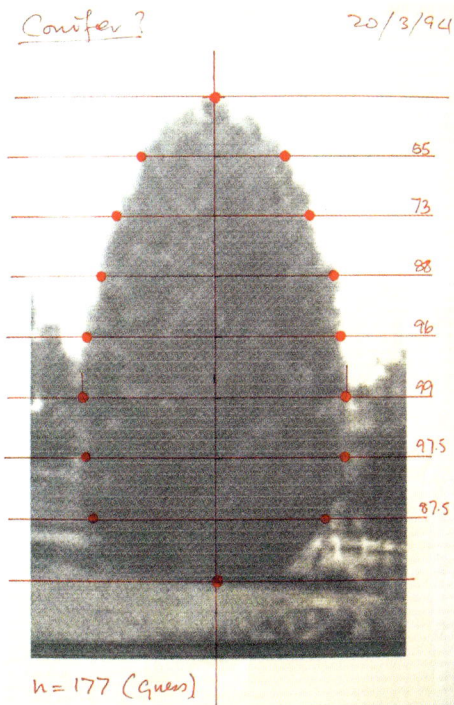

*Figure 10.42 Measurements of a conifer of unknown species*

*Figure 10.43* Tristania *measurements*

## 10.6 Tree outlines

I had wondered whether the profiles or outlines of trees might fit this pattern. One initial tree profile that I examined was that of the ubiquitous suburban conifer. The calculations gave a reasonable result, with a λ-value of 1.67 but a high mean lambda deviation of 16.2%. Trees are notorious for the difficulty of assessing where the ideal profile might be.

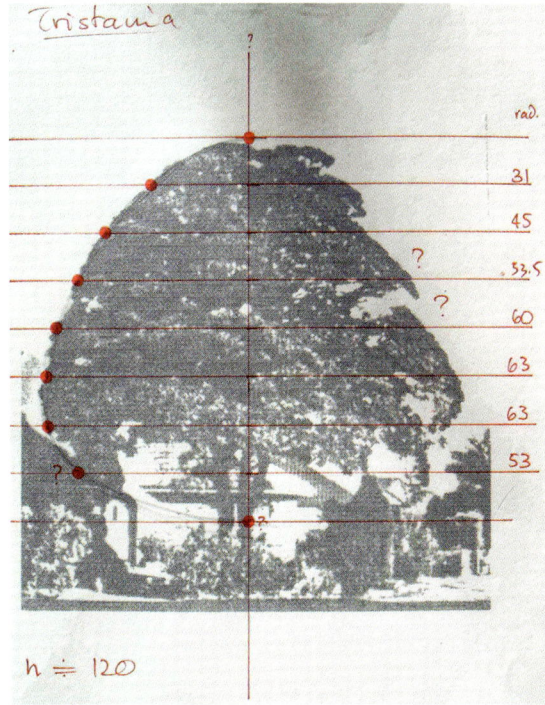

| tristania, | 20/3/1994 | | | | | | | | | |
|---|---|---|---|---|---|---|---|---|---|---|
| height X | 120 mm | | | | | | | | | |
| | measure | | | | | | | | | |
| level | dia(mm) | normalis | lambda | weight L | lambda de | ideal rad | | height | radius dev | |
| | D | D/T | L | W*L | abs dev | DI | DI*T | H | abs dev | % dev |
| Y | | | | | | | 0.0 | 120.0 | | |
| F | 31.00 | 0.5167 | 1.681 | 6.722 | 0.146 | 0.498 | 29.9 | 105.0 | 1.139 | 3.8158 |
| E | 45.00 | 0.7500 | 1.710 | 3.419 | 0.117 | 0.738 | 44.3 | 90.0 | 0.748 | 1.6903 |
| D | 53.50 | 0.8917 | 1.952 | 1.952 | 0.126 | 0.899 | 53.9 | 75.0 | 0.414 | 0.7688 |
| T | 60.00 | 1.0000 | | | | 1.000 | 60.0 | 60.0 | | |
| C | 63.00 | 1.0500 | 1.930 | 1.930 | 0.103 | 1.043 | 62.6 | 45.0 | 0.401 | 0.6404 |
| B | 63.00 | 1.0500 | 2.080 | 4.160 | 0.254 | 1.017 | 61.0 | 30.0 | 1.986 | 3.2547 |
| A | 53.00 | 0.8833 | 1.846 | 7.385 | 0.020 | 0.879 | 52.7 | 15.0 | 0.255 | 0.4832 |
| X | | | | | | | 0.0 | 0.0 | | |
| | | | | | | | | | | |
| means | | | 1.866 | 1.826 | 12.8 | | | | | 1.776 |
| | | | mean | weighted | %mean | | | | | % mean |
| | | | lambda | mean | lambda | | | | | radius |
| | | | | lambda | deviation | | | | | deviation |
| | | | | LM | MLD | | | | | MRD |

*Figure 10.44 Calculations of* Tristania *data*

*Figure 10.45 Overlay comparing theoretical (yellow form) with actual (red dots underneath)*
*Figure 10.46 The same tree after fourteen years in June 2008*

*Figure 10.47 Analysis of a sea urchin,* Amblyneustus pallidus
*Figure 10.48 Calculations of* Amblyneustus pallidus *and a representative but very approximate three-dimensional programmed model*

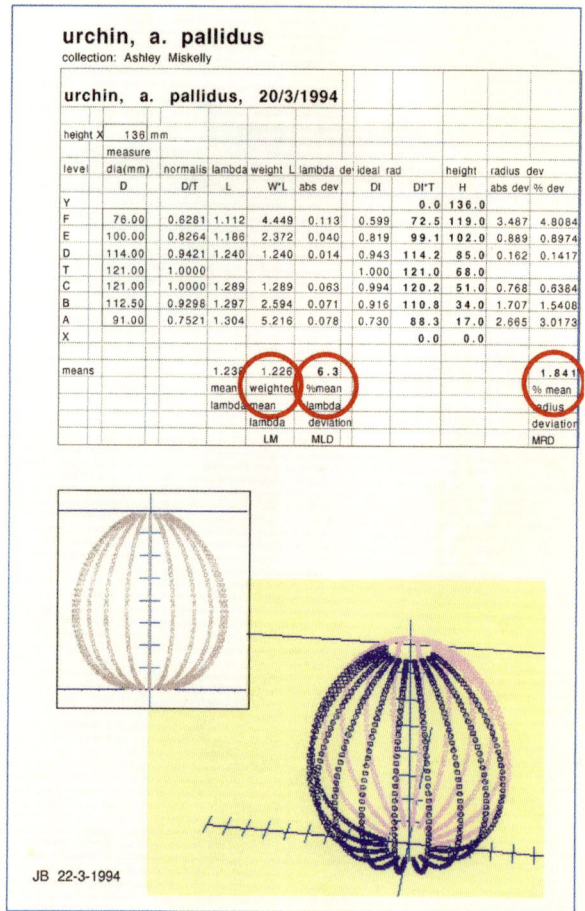

A. Pallidus    20/3/94

76
100
114
104
121
121
112.5
91

h = 136 mm

### urchin, a. pallidus
collection: Ashley Miskelly

**urchin, a. pallidus, 20/3/1994**

height X  136 mm

| level | measure dia(mm) D | normalis D/T | lambda L L | weight L W*L | lambda dev abs dev | ideal rad DI | DI*T | height H | radius dev abs dev | % dev |
|---|---|---|---|---|---|---|---|---|---|---|
| Y | | | | | | | 0.0 | 136.0 | | |
| F | 76.00 | 0.6281 | 1.112 | 4.449 | 0.113 | 0.599 | 72.5 | 119.0 | 3.487 | 4.8084 |
| E | 100.00 | 0.8264 | 1.186 | 2.372 | 0.040 | 0.819 | 99.1 | 102.0 | 0.869 | 0.8974 |
| D | 114.00 | 0.9421 | 1.240 | 1.240 | 0.014 | 0.943 | 114.2 | 85.0 | 0.162 | 0.1417 |
| T | 121.00 | 1.0000 | | | | 1.000 | 121.0 | 68.0 | | |
| C | 121.00 | 1.0000 | 1.289 | 1.289 | 0.063 | 0.994 | 120.2 | 51.0 | 0.768 | 0.6384 |
| B | 112.50 | 0.9298 | 1.297 | 2.594 | 0.071 | 0.916 | 110.8 | 34.0 | 1.707 | 1.5408 |
| A | 91.00 | 0.7521 | 1.304 | 5.216 | 0.078 | 0.730 | 88.3 | 17.0 | 2.665 | 3.0173 |
| X | | | | | | | 0.0 | 0.0 | | |
| means | | 1.238 mean lambda | 1.226 weighted mean lambda LM | 6.3 %mean lambda deviation MLD | | | | | 1.841 % mean radius deviation MRD | |

JB 22-3-1994

With the kind of profiles that many trees present it is easy to think that these too might meet the path curve, and the conifer was not too bad. There was a local *Tristania* or brush box that seemed to have had a constancy of form for many years and looked to be a good candidate for path curve comparison. Analysis gave a weighted mean λ-value of 1.826. Then an ideal profile was plotted and compared to the 'real thing' by doing an overlay. The result was not unreasonable. Some of the photos are a bit distorted as they were shots of older paper records. This was done in 1994 and it might be interesting to compare the profile with the tree fourteen years later. While a wall has appeared, the form of the tree itself is not *very* different.

## 10.7 The sea urchin

There is one more creature that I did some work on, and that was the sea urchin. I was lucky enough in my travels to come across Ashley Miskelly, author of *Sea Urchins of Australia and the Indo-Pacific*. His work with sea urchins I found extraordinary. His collection was beautiful, immaculate and well labelled. He allowed me to take a number of pictures of actual urchin tests (that is, the shell itself without spines) for a form analysis. I did a number of these but only record two species here.

I should mention that Nick Thomas, a colleague in this work, has his doubts about a simple path curve analysis here, for the urchin is usually re-curved at the mouth (bottom end), so the full form is not quite truly completely convex. While I respect his doubts, I include these samples as, superficially at least, the forms do respond to this analysis quite well.

The first sample is known as *Amblyneustus pallidus*. Its natural orientation (with the mouth down) results in a λ-value greater than one. On calculating I found it to be 1.23, with a 6% mean deviation. Another sea urchin, *Salmacis sphaeroides*, showed a λ-value of 1.486, but a λ-deviation that was not so good, though the radius deviation of only 5.3% was not so bad. Despite this the shell conformed visually quite well to an idealised profile as calculated and programmed. When the calculated profile was overlaid on the picture of the actual profile it corresponded fairly well.

*Figure 10.49 Analysis of* Salmacis sphaeroides
*Figure 10.50 Calculated profile path curves (with some other curves to indicate the spine markings of the urchin) of* Salmacis sphaeroides
*Figure 10.51 Overlay of calculated path on* Salmacis sphaeroides

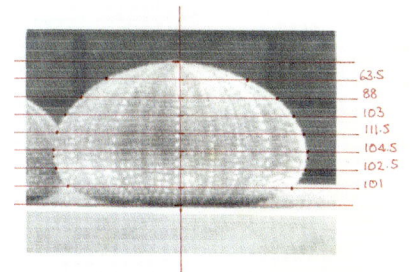

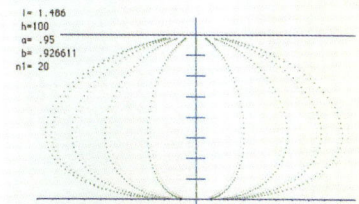

# 11. Concave Path Curves

There are two primal form gestures in both geometry and nature: the convex and the concave. The last chapter examined the egg or bud profile, a shape that has a convex, enclosing and embracing character. The opposite of this is the concave. Some work was done on this in section 9.2 on the airy and the watery vortex, as Edwards termed them. Here we shall look first at the airy vortex and where such a thing may be around us.

## 11.1 The grass tree and palm fronds

*Figure 11.1 Grass tree (*Xanthorea australis)
*Figure 11.2 A section through the frond base of a grass tree (after a bushfire)*

Grass trees are scattered around Australia. There are a number of species. They are mostly a single branch but can be two, occasionally three. They are a seemingly very simple plant and they

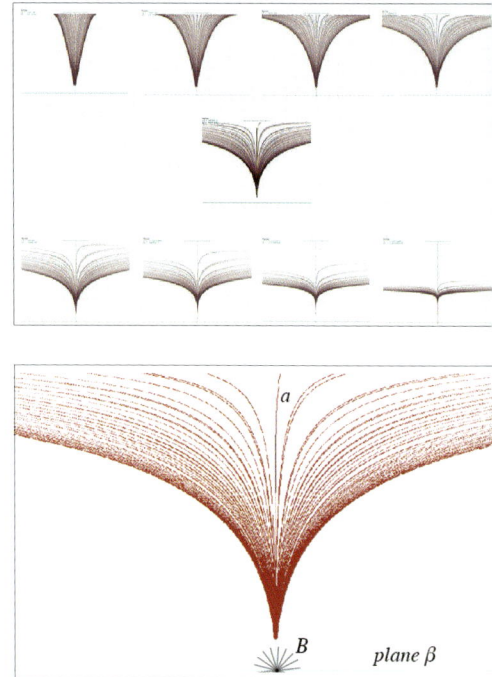

can be very old. The fronds, which are small quadrilaterals in section, grow from below and fan out. I have attempted to relate a vortex-like field to a tree with a single base, which had been trimmed (Figure 11.3). It looked as if it may have been burnt, not unusual in the Australian bush, but the essential habit of the raying fronds is typical.

The fields in Figure 11.4 are for a range of λ-values from –0.1 through –0.5 to –0.9. The form where λ = –0.5 is shown Figure 11.5 with the central line $a$ and the base plane β *(beta)*. The two intersect in point B. Then the computer generated fields of these path curves are matched to the fronds. The λ-value that seems to give a visual best fit is –0.5 (Figure 11.6). It is a curve fitting exercise that is far from exact. A lot of trial and error may be needed to get closer to this. The apex point B must be somewhere in the middle of the top part of the plant, and the horizontal line $z$ is assumed to be at infinity, but it is hard to be certain.

Is it a reasonable match? Certainly on some fronds it looks good. But it may be a catch-all approach. So I attempted to analyse in more detail the curve of a single palm frond, as the curved stem that carries the leaf blades away from the trunk have a firmly assertive curve. I selected one specific branch of the whole plant that was, more or less, in a plane perpendicular to our viewpoint and that cut through the central axis of the plant. How many points do we need on a curve to determine it as a function? A circle or a parabola for instance, is totally determined by three

*Figure 11.3 Grass tree in a nursery*
*Figure 11.4 Vortex profile range with λ-values from -0.1 to -0.9*
*Figure 11.5 Vortex with λ-value of 0*
*Figure 11.6 Overlay of vortex form over grass tree*

*Figure 11.7 Palm branches*
*Figure 11.8 Finding the geometric measure to determine the centre line*
*Figure 11.9 Calculated centre line*
*Figure 11.10 Finding base B*

points. The bud form needs four points – the top and bottom poles and two more.

First I estimated the centre line of palm. This is tricky. It could be estimated visually of course, but a calculation may give greater credibility. I assumed that all the fronds had more or less the same form factor ($\lambda$-value). This cannot actually be true, as with time the form changes (Edwards's later work on buds has shown this), but initially it is a reasonable approximation. This would mean that the measures on each side of the plant would be assumed to have the same $\lambda$-value which enables the geometric measure on a central axis to be calculated.

To do this we need three points on each of two curves, one on the left side and one on the right (Figure 11.8). The lines joining these (red) points cut any horizontal plane parallel to the base plane through B. If the form factor is the same on both left and right sides then the measures horizontally must start at the same point (on vertical axis $a$). The numbers in the horizontal plane will determine the position of this point. This is shown in Figure 11.9.

A similar procedure was adopted to find the position of the vortex apex, point B that determines the position of the horizontal plane $\beta$ on this vertical line (Figure 11.10). The position of the vertical line $a$ seemed plausible but the place for point B was much lower than I expected. To draw a vortex curve through the red points and through point B we need to take the two lower red points on the left hand side and finding multipliers both horizontally and vertically (two red

Vortex grid

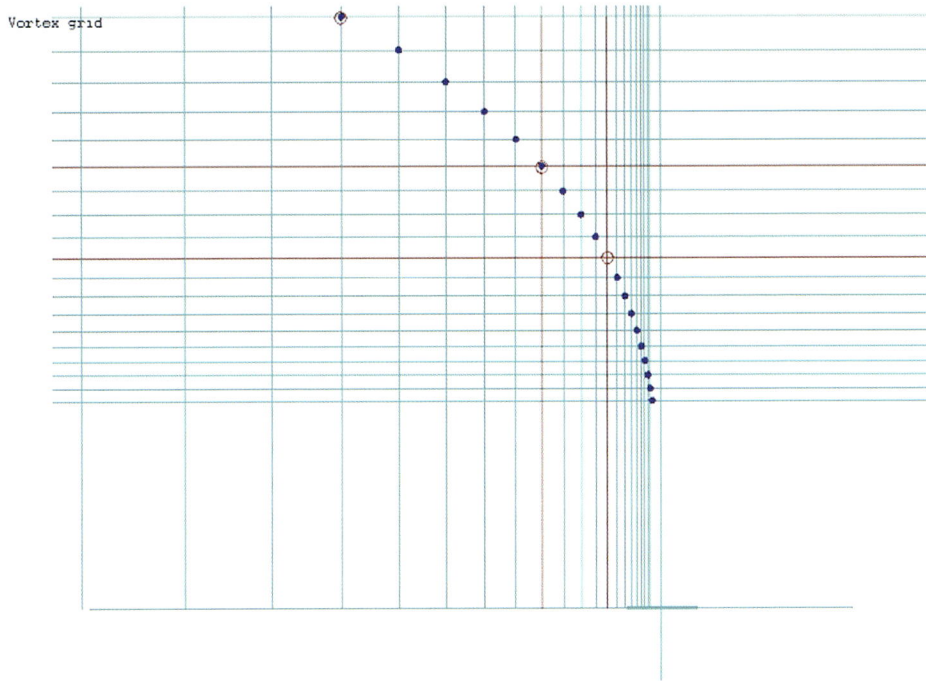

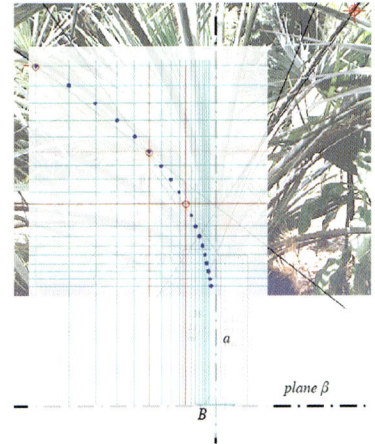

horizontal lines and two red vertical lines), referring to point *B* as the origin of both geometric measures.

These gave a twofold geometric scale measure of 2.238 horizontally and 1.26 vertically (Figure 11.11). This was then overlaid and positioned on the image of the palm fronds. To find the vortex profile plot on consecutive pairs of horizontal and vertical lines, starting at one of the known points. Selecting the lower red data point on the left and moving across the rectangles creates a curve, and join the dots. The left hand plotted curve came out very well, mirroring closely the underside of the frond. Using Edwards's method, the calculated λ-value was found to be –0.714.

Using this λ-value the whole field of curves can be plotted, resulting in a family of nested vortices. I plotted this field of curves by computer. Overlaying this on the palm shows that many of the fronds do not fit as well as our first stem did (Figure 11.13). This may have been due to assuming all fronds were exhibiting the same λ-value. To be convincing, this analysis would obviously need to be done many times with many plants. No doubt the assumptions would be progressively refined in the light of experience.

The gesture of the airy vortex is found in the fundamental branching structure of plant forms. Does such a field guide the branches and stems as they grow? A lot more work needs to be done to verify this. On a larger scale this form is also seen in cyclones, hurricanes, typhoons and waterspouts. Again this is a vast field for further work.

*Figure 11.15 Winter treescape, England, shows the two formative aspects*
*Figure 11.16 Dragon tree [Daniella]*

## 11.2 The interplay of concave and convex

These budding, convex formative tendency and the branching, concave tendency often work together in nature. The canopy of the dragon tree, the outline of the oak tree, the umbrella of the eucalyptus are all reasonably well-defined convex profiles, and at the same time the concave, raying aspect can be seen in the branching. Something can be seen in the dragon tree with its tightly well-formed canopy and the branches in a continual doubling as they grow.

Is it some special intuition that leads many gardeners to plant alternating bud forms (ovals) and vortex forms (dart)? And what about the classic dart and oval (or egg and tongue) design, now relegated to cornices near the ceiling of older mansions?

Imagine my astonishment when I realised that both the primal forms, the convex budding and the concave raying, are inherent in the same path curve field. I became aware of this connection through Lawrence Edwards. In Figure 11.20 one point is at infinity and the line opposite is perpendicular to the other two. The vortex profiles from $X$ can just as easily be sourced from $Y$. In fact depending upon how the points in the field are joined there can be a multitude of profiles. These curves are generated from

*Figure 11.17 A garden with alternate forms of plants in Totnes, Devon, England*

the intersection of rays, so this always gives the possibility of *two* sets of curves, depending which diagonal of the little rectangles is joined.

Note that the material substance *is* the shape or fill of the plant, but that the forming *of* it as such is invisible, for the curves *surround* it and define its limits. It is just as important to see the invisible (we do it through geometry, through ideas) as it is the visible (through the senses). Is it conceivable that *real* ideas (the archetype?) brought about the cabbage, the beetroot or apple?

When one point of the triangle is at infinity then the path curve profiles become bilaterally symmetric. We can find both types of curves in this field – the green convex and the red concave profile (Figure 11.21). It is just a question of how many little quadrilaterals are crossed each time and in which direction. (For instance, the red curve is made by crossing two lines that come from the top and one line from the bottom.)

*Figure 11.18 Alternate planting along a main road in Sydney*
*Figure 11.19 The classical frontage of the New South Wales Art Gallery in Sydney shows the dart and oval design in its cornice*

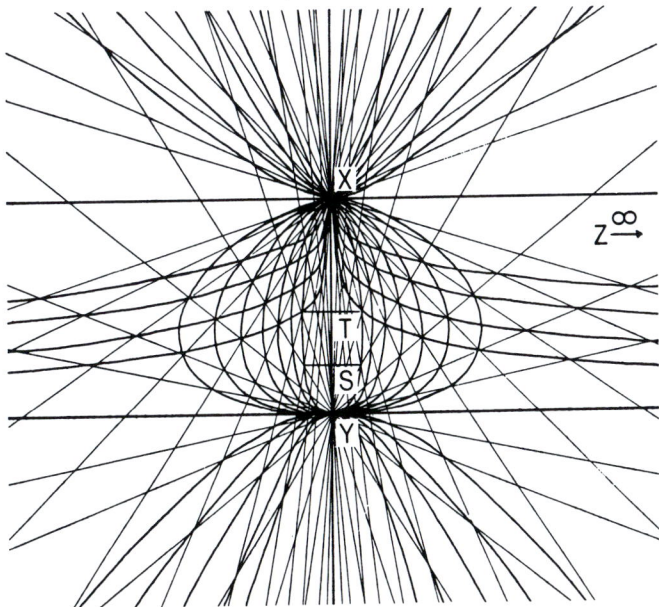

*Figure 11.20 The bud form and vortex form arising out of the same field (from Edwards,* The Vortex of Life)
*Figure 11.21 Special case of contra fields, both bud and vortex shapes together*

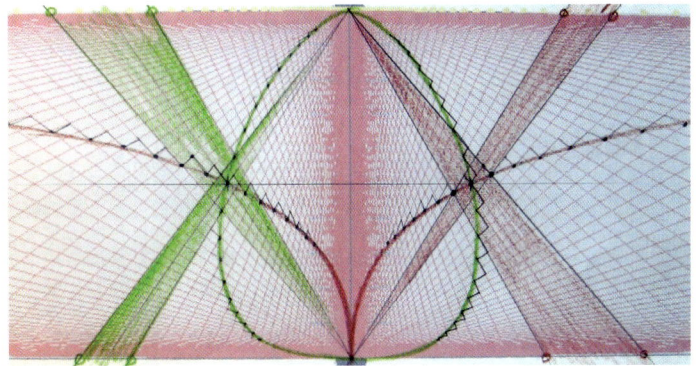

# 12. Form in the Mineral World

## 12.1 The fields of an all-real tetrahedron

In section 9.3 we considered the simplest three-dimensional element, the tetrahedron, and considered the forms that are generated by it. Although this was examined previously, until now we have been looking at profiles of spatial forms, in other words at two-dimensional curves. The curves generated by the tetrahedron are actually spatial curves, a three-dimensional form. We shall begin by looking at the normal, everyday tetrahedron, the all-real one.

We recall that the tetrahedron has four points (vertices), six lines (edges) and four planes (surfaces), and we can imagine the tetrahedron made up of or defined by points, lines or planes. And in section 3.4 we saw the interdependence of these elements.

*Figure 12.1 A tetrahedron made of points, lines and planes*
*Figure 12.2 Movement of points and planes in lines of the tetrahedron*

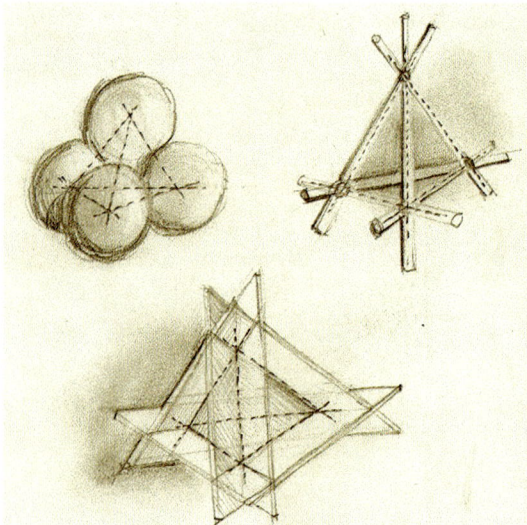

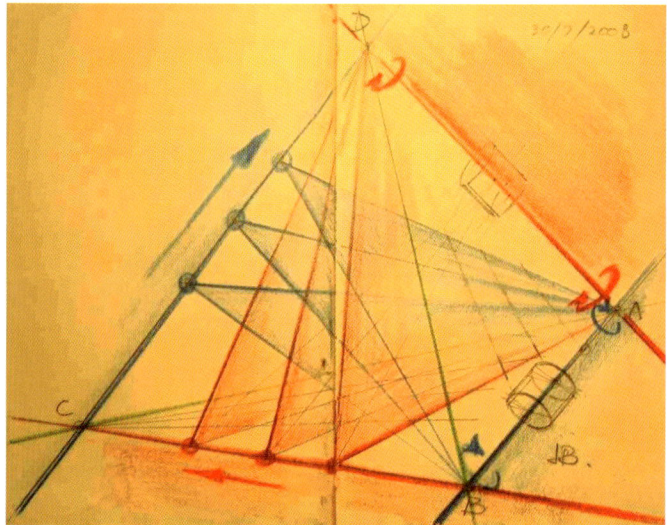

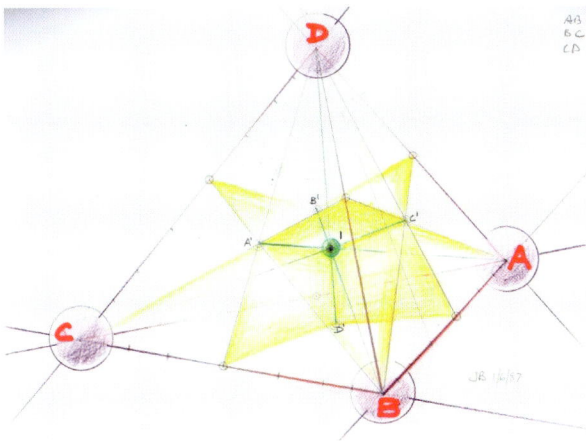

Figure 12.3

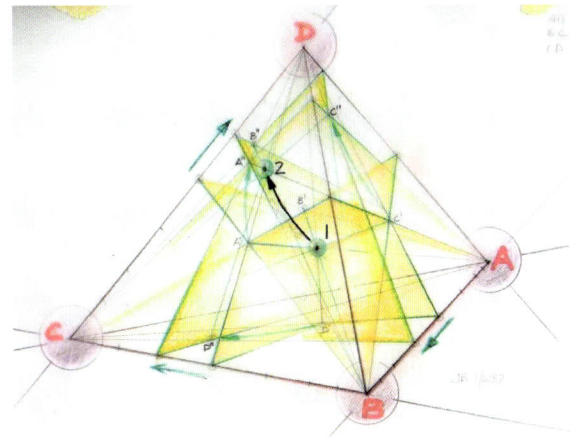

Figure 12.4

Just as with the triangle we allowed the line to transform in growth measure, so now with the tetrahedron we have planes, points and lines transforming in growth measure. In Figure 12.2 the blue points are moving in a growth measure along their line on the right. The blue planes are swinging in their line to the left. These two activities are coupled. This should never be forgotten. The two blue skew lines cannot behave but in an interconnected manner. A rotation of planes around a line is always linked to a translation of points in its skew line.

When space transforms into itself we arrive at *forms*. Here the point-line-plane triple moves to the rhythms along and about all six lines. The details can be found in Edwards, *The Vortex of Life*, Appendix 2. I shall summarise it in the following drawings.

Consider a general point, *1*, that might be situated somewhere in the middle of the tetrahedron *ABCD* (Figure 12.3). Note that this point is immediately connected with lines which join to each tetrahedron vertex *A*, *B*, *C* and *D*, as well as with planes in the lines *AB*, *BC*, *CD*, etc.

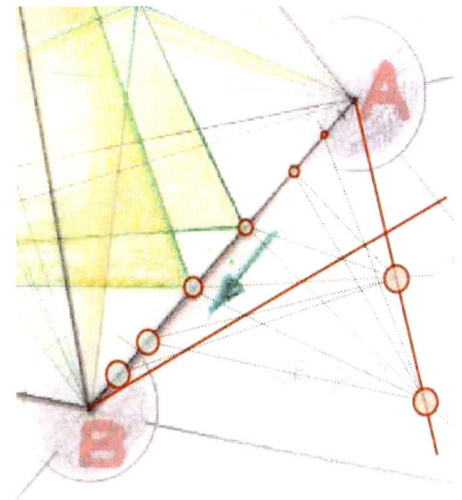

Figure 12.5

Move the point *1* to a new position, point *2* (Figure 12.4). This action causes all the lines and plane just mentioned to also move (green arrows) keeping the original tetrahedron intact. Note the new positions of the yellow planes and the green lines. The positions of all the points along three lines of the tetrahedron have now been determined. In each case we can extend the measure, as it is totally set.

We simply create a growth measure with these two known points and the two fixed points, *A* and *B*, for instance (Figure 12.5). This is because any four points determine such a measure.

Similarly we can also draw the measures along the other two determining lines, line *BC* and line *CD* of the base tetrahedron. The entire motion is now determined. This is so because the step

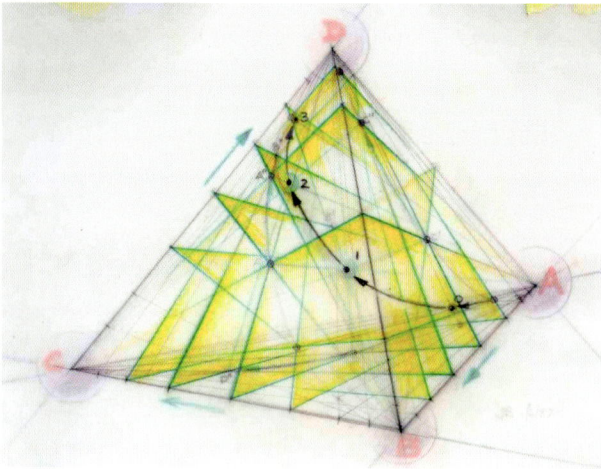

Figure 12.6 Additional points on the path curve
Figure 12.7 Path curve within the real tetrahedron

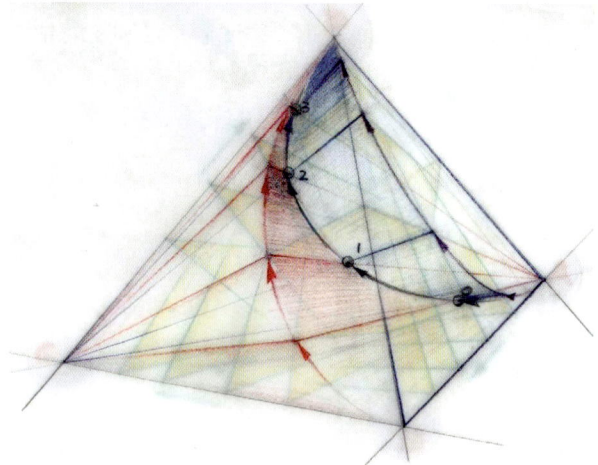

taken *along* each edge, and *around* each edge, has dictated the growth measures along and around these edges. Each of the three measures can be different, but these three, once set, determine the measures on *all* the remaining edges. We can now draw in as many points on the curve as we want to. Some are added in Figure 12.6.

In Figure 12.7 the path curves followed in the two planes *ABD* and *BCD* are emphasised. The intersection of the rays from *A* with the curve in plane *BCD* is one of the surfaces the curve *0, 1, 2, 3* must follow. The whole curve, of course, extends outside the tetrahedron, far beyond *A* and *D*, traversing all space, though neither the 'inside' part of the curve, nor the 'outside' part ever reach *A* or *D*.

This is *one* path curve for an all-real tetrahedron. This curve can stand alone, but whole families can be taken too – these are then

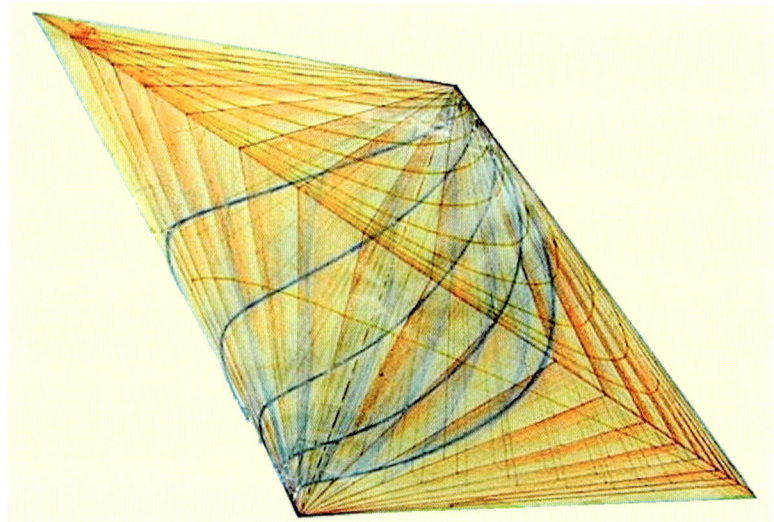

Figure 12.8 Path curve surfaces within an all-real tetrahedron

path-curve surfaces. I have tried to indicate that in Figure 12.8, though only the curves 'inside' the tetrahedron are shown.

It is not easy to visualise or to draw. I found it helped to make a model of a regular tetrahedron to affirm how it all was. The six lines can be thought of as three pairs of skew lines that are perpendicular to each other (Figure 12.9). Linking the opposite edges were thin steel threaded rods. In an early model (Figure 12.10) these bent somewhat, so for the next model I used slightly stronger rods. The length of each edge of the model was 1 metre (3 ft). To fix the centre of the system I made a small wooden block (about 35 mm, 1.5 in square) through which the rods passed (Figure 12.11). Since I did not want to cut the rods (so as to maintain strength and alignment) the lines (or rods) were deliberately offset. This compromise was an approximation. The next job was to fix all the four outer points rigidly. These joints were tricky (Figure 12.12).

Now the task was to construct three sets of interacting planes. I simply marked out a growth measure along each of the two rods

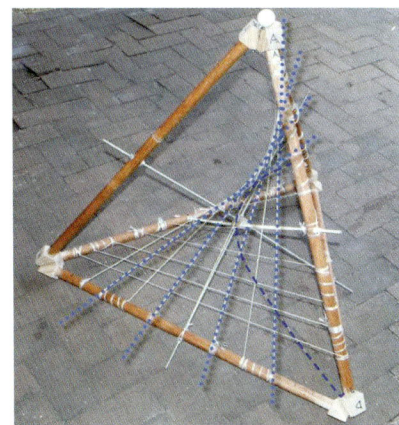

*Figure 12.9 Two skew and perpendicular lines; one of three pairs forming a regular tetrahedron*
*Figure 12.10 An early model of the regular tetrahedron*
*Figure 12.11 Centre block of model*
*Figure 12.12 End point joints*
*Figure 12.13 Model with one string surface*

*Figure 12.14 Model with three coloured string surfaces*

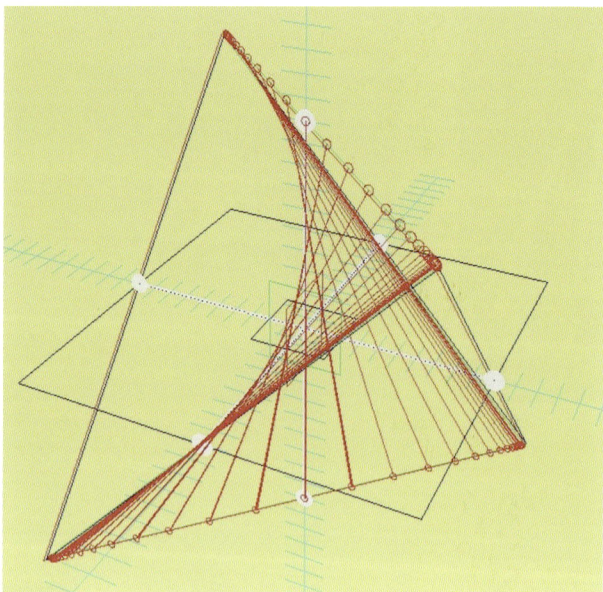

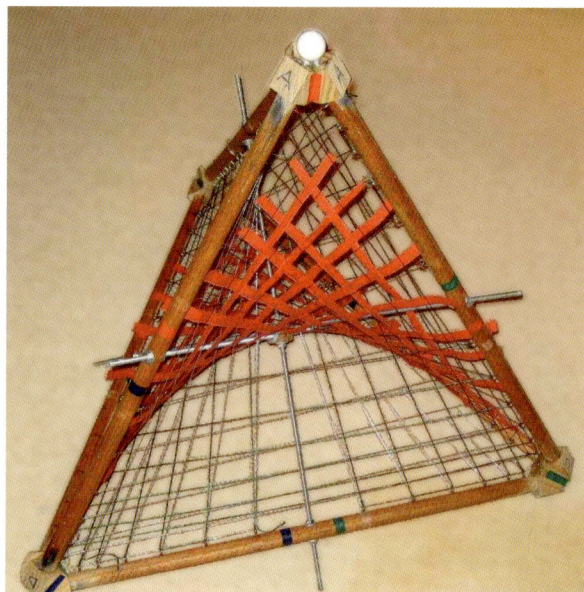

*Figure 12.15 A single ruled surface*
*Figure 12.16 Saddle surface*

*Figure 12.17 Three interpenetrating saddle surfaces*
*Figure 12.18 The centre where the curves are mutually perpendicular*

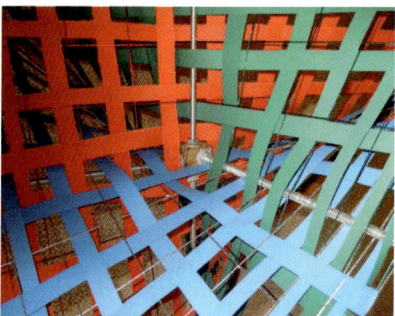

and joined with cord or string pair-wise across the system (Figures 12.13 and 12.14). Despite each cord being taut, and hence straight, the path surfaces appear as curved, in fact a doubly curved sheet with both concavity and convexity. It is a kind of symmetrical 'saddle form' pinned to the tetrahedron like a spreading shade cloth or tent at the four corners of the tetrahedron. Figure 12.15 only shows *one* surface; there are in fact *three* such surfaces. Each surface is really defined by two sets of ruled lines. To emphasise the surfaces I have woven a red ribbon into the mesh (Figure 12.16). At the centre, at the middle of the saddle, we see how the surface tends to flatten out. Note again that the surface includes the ruled lines from four of the tetrahedron's edges, and not just two of them.

If now two further surfaces are added (say green and blue), perpendicular to the red surface at the centre, then another curious feature emerges. The two surfaces both twist and the merge. They twist away from each other in the middle. They merge in the tetrahedrons lines (Figure 12.17). We go from a central rigid three-dimensional Cartesian framework through three intersecting saddle forms to the edges of a regular tetrahedron.

## 12.2 The infinitely large all-real tetrahedron

Returning for a moment to the intersecting lines towards the centre of the tetrahedron we can see a cube-like structure emerge. Within this are two tetrahedra – as with any cube. These are emphasised with the red cord and green cord (Figure 12.19). The cube-like structure, however, is not quite a cube. The cords or lines forming the top face are not quite in a plane (Figure 12.20). The larger the tetrahedron becomes, the flatter this faces becomes and the more cube-like the solid. When the tetrahedron is infinitely large it actually becomes a cube.

Two things happen when the tetrahedron becomes infinitely large. The three lines remain perpendicular to each other as we move away from the 'centre', and the growth measure along the infinitely distant lines of the tetrahedron becomes a step measure – so that the lines appear as a regular rectangular grid. Figure 12.21 tries to show this. In the centre we see a rectangular prism. Such an entity could be repeated indefinitely throughout the infinite tetrahedral volume.

The equally-sized repeated form is akin to crystal structure. We can try and imagine the flow or stream of the raying lines of this infinitely large tetrahedron, the streams coming from different edges of the tetrahedron, moving in different directions, but in similar rhythm. There would be a field similar in nature to standing waves, and the regular nodes creating structure into which the mineral, crystal world coalesces. This is obviously not an explanation of the physics of crystal forms, but we can feel a certain similarity in character between the crystal structure and the field emerging from the infinite all-real tetrahedron.

*Figure 12.19 Two tetrahedra within the central lattice-cube*
*Figure 12.20 The central cube is not actually a cube*
*Figure 12.21 Surfaces of path curves of an infinitely large tetrahedron*

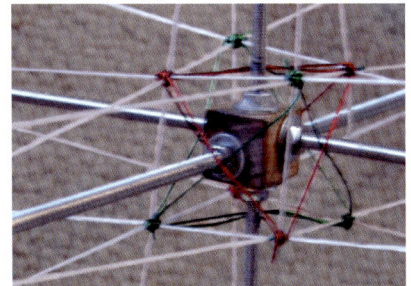
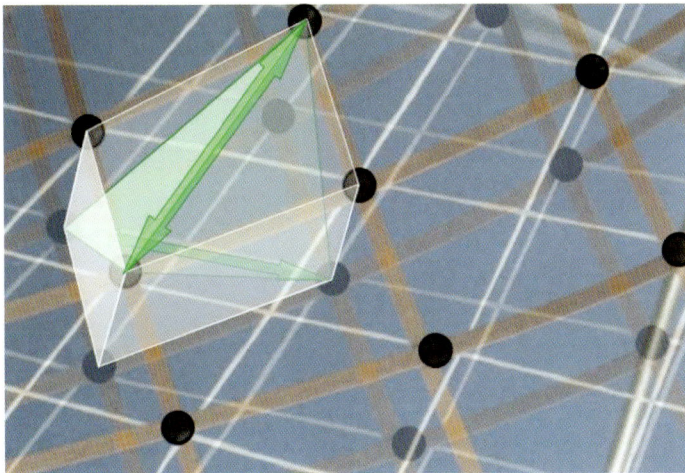
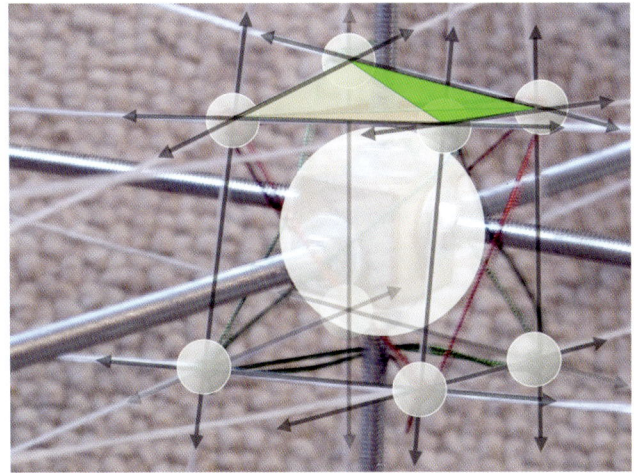

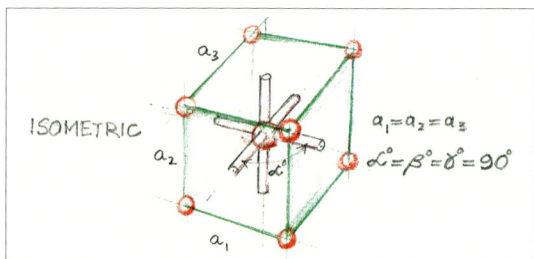

Figure 12.22 Cubic or isometric crystal structure

$a_1 = a_2 = a_3$

$\alpha = \beta = \gamma = 90°$

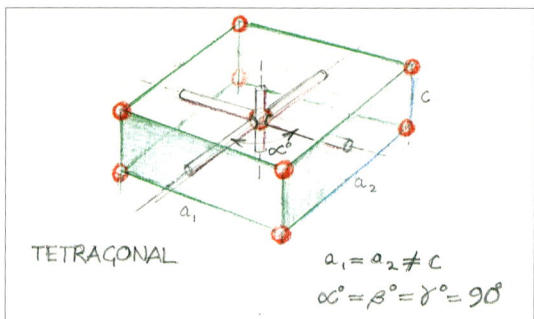

Figure 12.23 Tetragonal crystal structures

$a_1 = a_2 \neq C$

$\alpha° = \beta° = \gamma° = 90°$

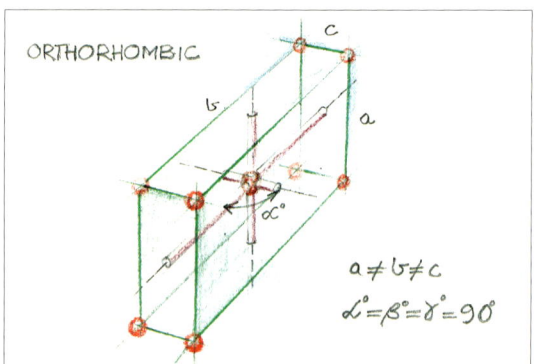

Figure 12.24 Orthorhombic crystal structures

$a \neq b \neq c$

$\alpha° = \beta° = \gamma° = 90°$

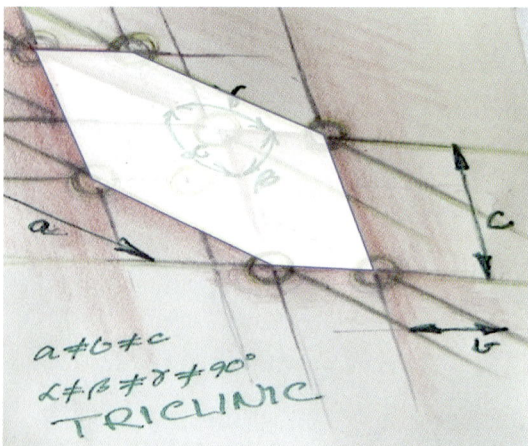

Figure 12.25 Triclinic crystal structure

$a \neq b \neq c$

$\alpha \neq \beta \neq \gamma \neq 90°$

TRICLINIC

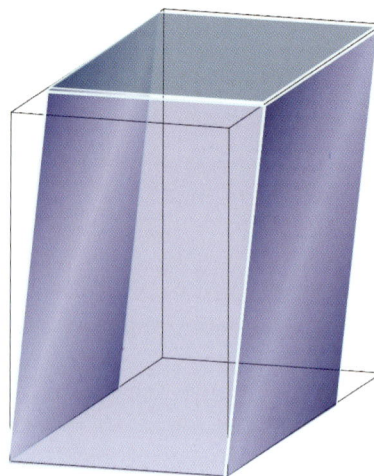

Figure 12.26 Monoclinic crystal structure

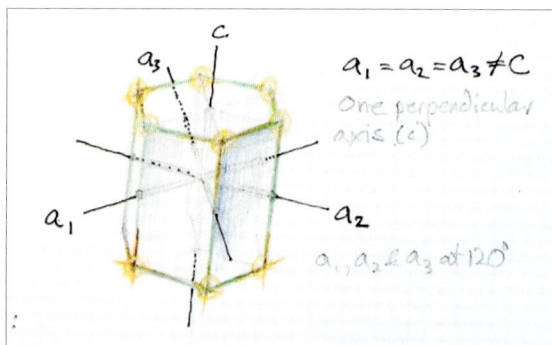

Figure 12.27 Hexagonal crystal structure

$a_1 = a_2 = a_3 \neq C$

One perpendicular axis (c)

$a_1, a_2 \& a_3$ at 120°

Figure 12.28 Trigonal crystal structure

## 12.3 Crystal structures

There are six or seven crystal systems (the difference arises as the hexagonal and trigonal are very alike, so some authorities class them as one system). The first three are all based on the right angle, that is, their axes are all mutually perpendicular. They are the cubic or *isometric,* the *tetragonal* and the *orthorhombic.* The difference in these three, is that the side lengths can be varied. The isometric has all sides the same length (Figure 12.22). The tetragonal has two sides equal and the third can be larger or smaller than these (Figure 12.23). The orthorhombic has sides of three different lengths (Figure 12.24).

A further system is the *triclinic* (Figure 12.25). Adjacent sides are not equal and the angles are not equal to each other and are not right-angled. This is the most general case (while the isometric is the most regular). The *monoclinic* case is where adjacent sides are not equal, the two base angles are right angles but the third is not perpendicular to the base (Figure 12.26). Most crystal structures are monoclinic. The last two systems are the *hexagonal* and the *trigonal.* The hexagonal is basically sixfold about a central axis, while the trigonal is threefold also about a central axis (Figures 12.27 and 12.28).

In 1784 the Abbé (René Just) Haüy, the French father of modern crystallography, claimed the external regularity of these forms 'is due to symmetry at the microscopic (atomic and molecular) level'. So whatever the particular crystal appears like externally it is still constituted of identical unit cells that simply repeated themselves. Yet it is astonishing that the aggregates we call crystals have a macro form often clearly recognisable as a particular material. Stacking itself does not entirely explain the overall form.

The infinitely large tetrahedron could in a sense be the basis for cubic and rectangular prismatic forms. The planes of such forms would be related to the six infinitely distant edges of the tetrahedron and, as we saw, the growth measure on these edges turns into an equal step measure.

There are many examples of this structure. One is simple table salt. My favourite is iron pyrites, commonly called 'fools gold'. This mineral can form the most perfect looking cubes and rectangular prisms. Fluorite, garnet, galena and many more are cubic. Natural planes allow fluorite, for instance, to be cleaved into the octahedral form. The wooden block model of cubic unit cells shows how these may be stacked to give an octahedron – an obviously quite different form to the cube.

Another form that can be built of these unit cubes is the rhombic dodecahedron. The garnet often matches this form. Another fine example of cubic crystal form is lead sulfide, or galena. The sample

*Figures 12.29 and 12.30 Iron pyrites crystals*
*Figure 12.31 Fluoride cleaved as an octahedron*

here is like a little landscape built of cubic bricks. This material breaks up easily and leaves multiple rectangular surfaces.

The tetragonal (only two side lengths are the same and all angles are right angles) differs from the isometric as only two of the rhythms of the infinitely large tetrahedron would be the same, the third being different. Examples of this structure include rutile, cassiterite, and idocrase.

In the orthorhombic crystal system none of the sides are of equal length, but all the angles are still right angles. On the great tetrahedron there would be three different step measures on opposite skew line pairs. Examples of this structure include barite and olivine.

One peculiar characteristic of the path surfaces that we have ignored when imagining the infinitely large tetrahedron is their orientation. As the tetrahedron grows the curved saddle surfaces flatten out until they become perfectly flat when the tetrahedron is infinitely large. But the lines that these flat plains connect are mutually perpendicular, so the surfaces actually have to twist through 90° from one side of the tetrahedron to the other, and yet they remain flat. Now there's a twist!

*Figure 12.32 Octahedron formed of cubes; Figure 12.33 Glass model of a rhombic dodecahedron [Christel Post]; Figure 12.34 A garnet; Figure 12.35 Galena (lead sulfide)*

The monoclinic only has two axes at right angles, the third at some other angle. Examples of this include gypsum, epidote, lazulite, mica, orthoclase feldspar and saleeite. The triclinic has all different angles and no right angles. Only 7 per cent of crystal forms conform to this structure. Examples include kyanite, axinite, microcline, wollastonite and rhodonoite. With regard to the tetrahedron, it is no longer regular.

The remaining two forms elude me as to any relation to a tetrahedral basis. This could mean it simply is not there. Quartz is a wonderful example of the hexagonal structure. The trigonal structure, similar in nature to the hexagonal, has an astonishing example in the tourmaline.

*Figure 12.36 Orthoclase feldspar, monoclinic structure [Albert Chapman collection, Australian Museum, Sydney]*
*Figure 12.37 Saleeite, monoclinic structure [Albert Chapman collection, Australian Museum, Sydney]*

*Figure 12.38 Microcline feldspar, triclinic structure [Albert Chapman collection, Australian Museum, Sydney]*
*Figure 12.39 Smoky quartz, hexagonal structure*
*Figure 12.40 Tourmaline, trigonal structure*

# 13. Form in the Plant World

## 13.1 The semi-imaginary tetrahedron

We saw in Chapter 10 that the plant world (and some aspects of the animal world) exhibits a congruence with path curves. This was the focus of Lawrence Edwards's early work, but he went on to apply this geometrical method to forms in space.

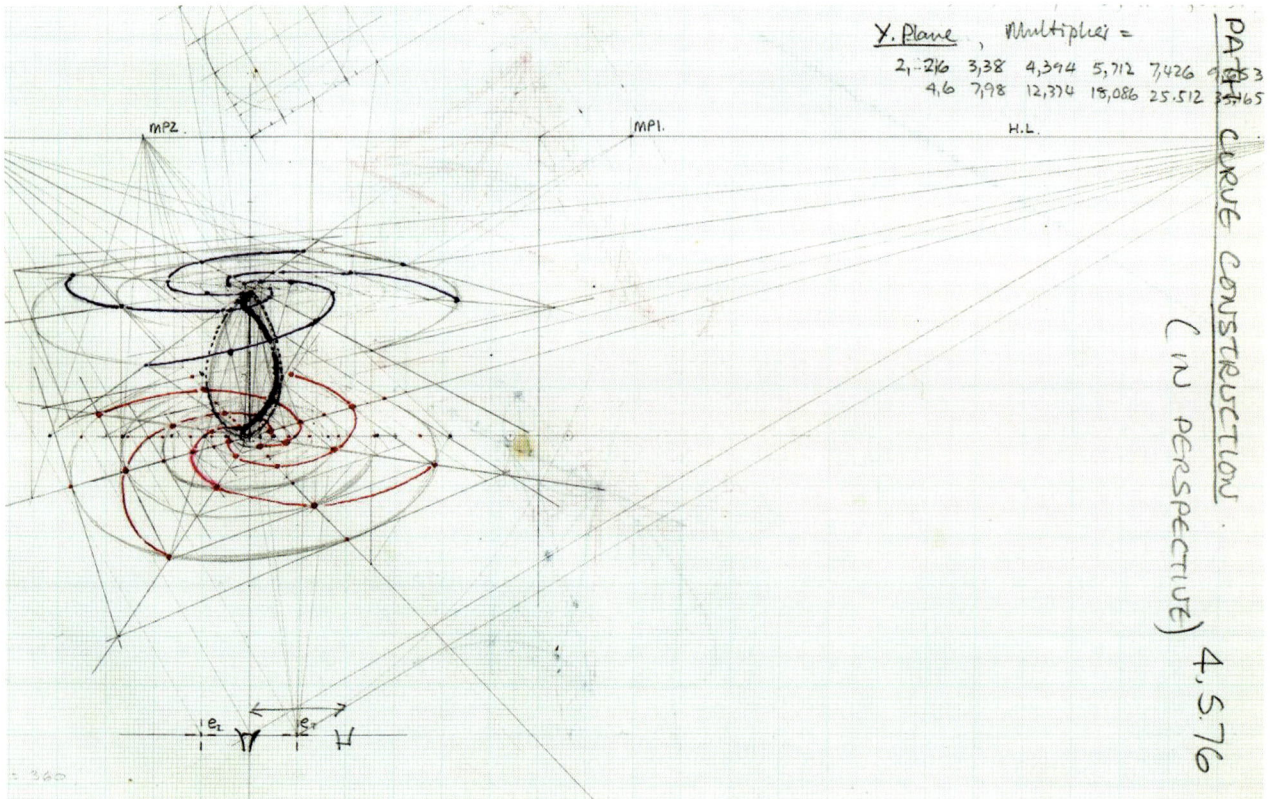

*Figure 13.1*

The tetrahedron required for the formation of these forms is the semi-imaginary tetrahedron, not the all-real one that we looked at in the last chapter. We examined this semi-imaginary case in section 9.2 on the vortex. That semi-imaginary tetrahedron has two real lines (one vertical, and one horizontal at infinity) and two real points (on the vertical line). The rest of the elements - two of the four planes, two of the four points and four of the six lines – are imaginary; that is, in motion.

On the vertical, real-line element is a growth measure, not the equal sized step measure we had for the infinitely large tetrahedron, but the equal measure on the line at infinity is due to a circling of equal angles around the local vertical real line. These rhythms determine how the point/line/plane triples behave. They are the path curves that build the fields of forms in nature that we are interested in.

As mentioned earlier, I began this work when Lawrence Edwards was in Australia in 1976. Coming from an engineering background I wondered whether it was possible to find a system that could deal with the curves of life. I still have my first perspective path curve drawing for this special tetrahedron, from that time (Figure 13.1). This was drawn from a point-wise aspect, that being the easiest to manage. Notice that the two sets of path curves in the top plane (blue spirals) and the bottom plane (red spirals) run counter clockwise to each other, the blue clockwise and the red anticlockwise (see Chapter 10 and Figures 10.18 and 10.19). These fields are proper to the planes through top and bottom points and through the line at infinity.

Having set up this tetrahedron, we can now use two of these fields in the two planes set an arbitrary distance apart and rotating in opposite directions around the real vertical axis. Figure 13.2 puts a single blue spiral in the top plane and a single red spiral in the bottom plane. Both are clockwise when viewed

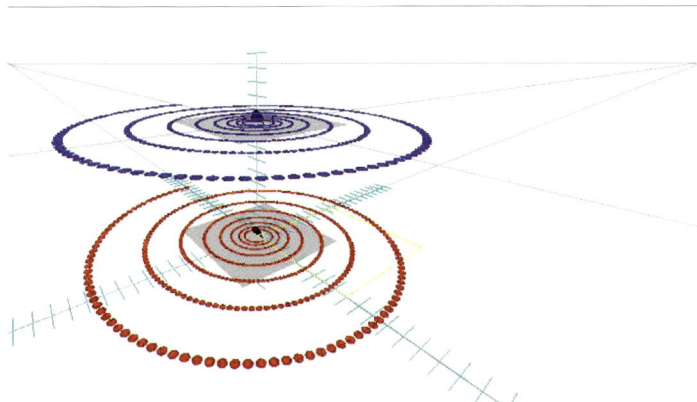

*Figure 13.2*

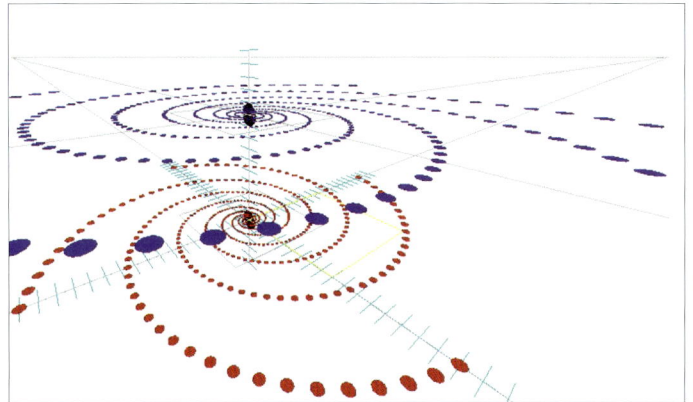

*Figure 13.3*

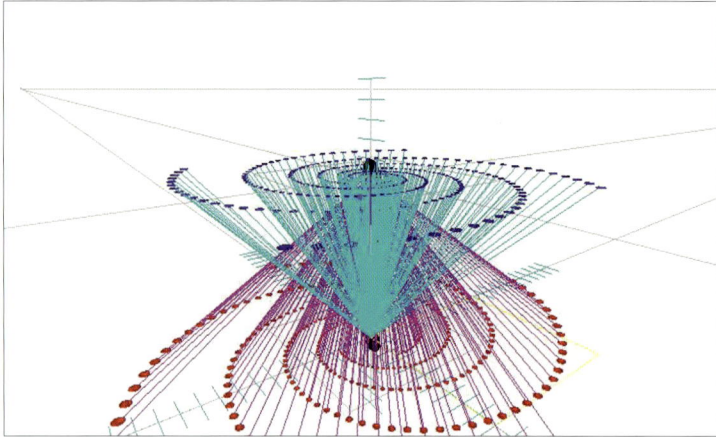

*Figure 13.4*

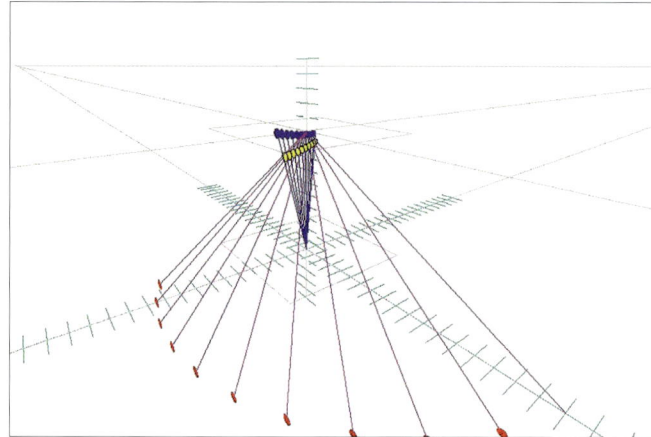

*Figure 13.5*

from above. These two curves are the result of point/line pairs in action (see Chapter 8, Figure 8.20 for the process). The next step is to make the red lower spirals rotate in the opposite direction and to multiply the number of spirals in both these directions (Figure 13.3).

There must also be the spiral cones of lines (and planes) in the two real points. The earlier drawing barely hints at it, but it is indicated clearly in Figure 13.4. Here we have the two-dimensional spirals in the two planes, as well as the spirals in the two points, which of course are also two-dimensional. Now we can see how these sets of elements interact.

As the original drawing only had one path curve, I set the computer to only plot one too. For this I used just one spiral in the top plane and one in the bottom plane and their counterparts in the two fixed points. One way to see the curves is to look where the lines of the cones interact. This is shown in

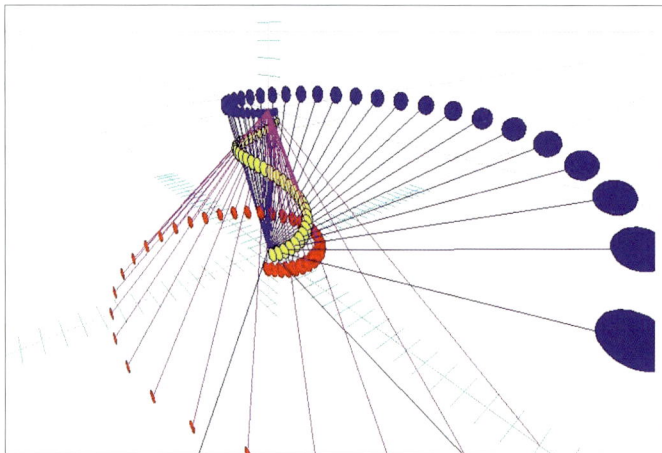

*Figure 13.6 Continuation of the path curve form*

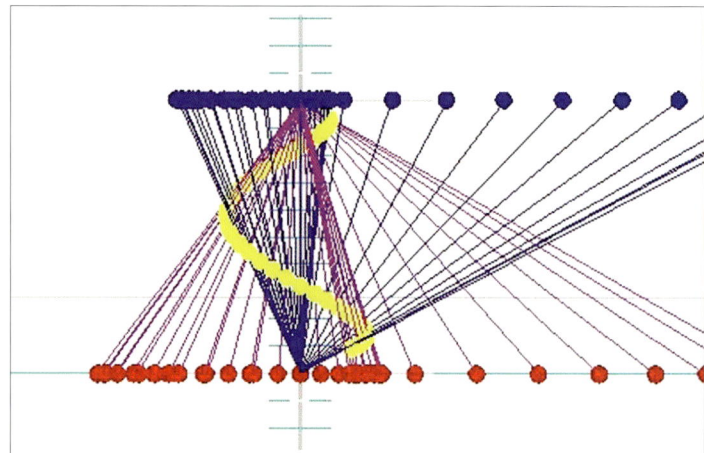

*Figure 13.7 Elevation*

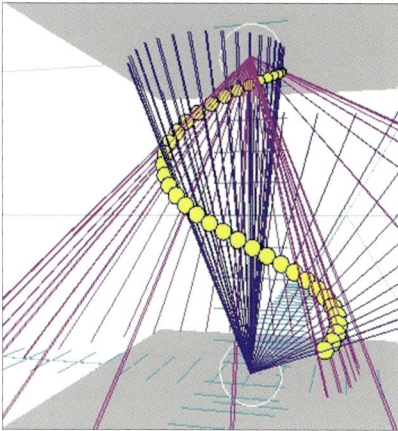

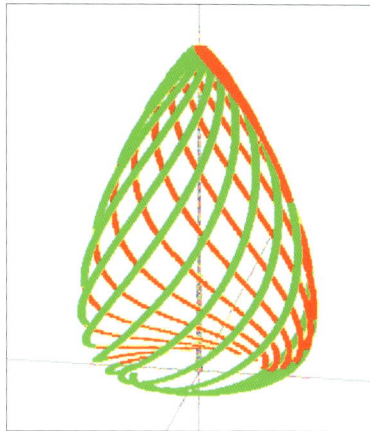

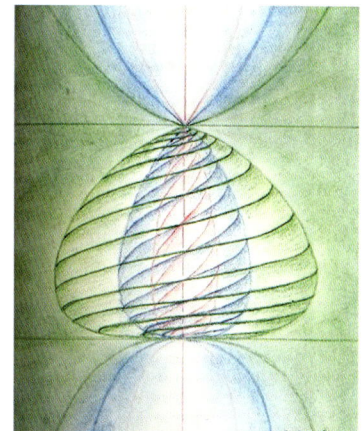

Figure 13.5 where we track the path of the intersection of pairs of lines (yellow dots). This is part of one of many possible path curves. The curve continues around the central vertical line (Figures 13.6 to 13.8).

Originally I drew four sets of spirals. What would the complex appear like with four or more such space curves? The crude piece of programming produced Figure 13.9 that shows fifteen identical curves displaced around the central vertical line. These curves sweep out an egg-shaped volume around this axis. Viewed from above it has a circular cross-section at all levels. The plant world predominantly displays such a circular horizontal cross-section. Are these cross-sections merely part of such a rotating S-curve, such a spinning path curve?

## 13.2 Lambda, epsilon and nodal rhythms

The overall form of the egg or bud-shape can vary as we saw in section 10.4. Edwards used $\lambda$ *(lambda)* as the factor to describe these forms. Figure 13.12 shows these forms now arising out of the rotation of spirals in space, and not by any arbitrary cross-sections as before.

*Figure 13.8 Close-up of intersecting cones*
*Figure 13.9 Multiple path curves*
*Figure 13.10 Egg-shaped path curves*
[*Olive Whicher,* Plant between Sun and Earth]
*Figure 13.11 A stained glass rendering that I made in the early days of my work*

*Figure 13.12 $\lambda$ (lambda) variations (from left: 0.5, 1.5, 5.0)*
*Figure 13.13 $\varepsilon$ (epsilon) variations (from very high value to low value approaching zero)*

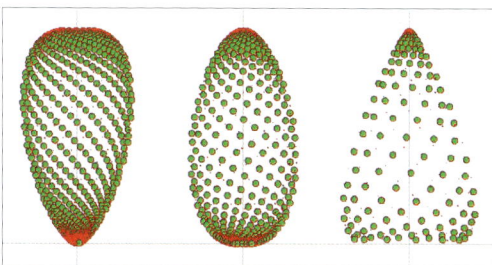

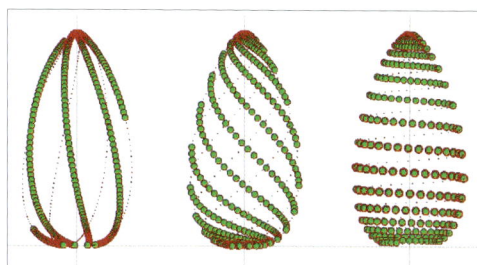

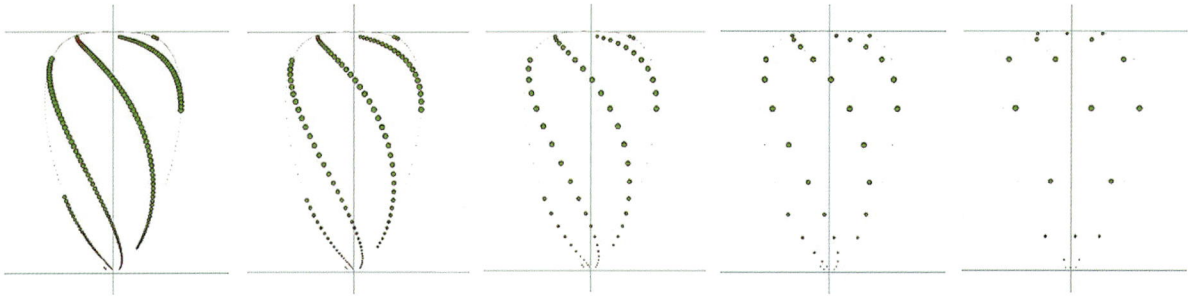

*Figure 13.14 The nodal rhythms or measure along the spiral curve (from 2° to 32° steps)*

*Figure 13.15 Computer plot produced by Lawrence Edwards in 1982*
*Figure 13.16 Early computer plot on a Casio hand-held calculator*

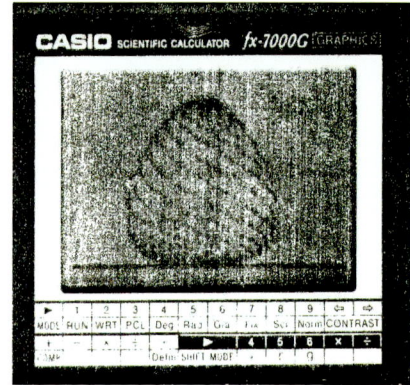

Another factor is the way the slope of the curves varies. Edwards called this ε *(epsilon)*. The ε-values range from zero (where the curve simply circumnavigates the form without slope) to infinity (where the curve simply goes straight up of down). I have not come across the zero-value, horizontal in the plant world, and have only seen the vertical in the sea urchin (hardly a plant of course). The plant world seems to demonstrate almost every possible value in between. For ε-value the curve can go in either direction.

There is still another factor that can vary, and I have called this the *nodal rhythm*, though it might more properly be called the *measure* along the spiral curve. This indicates how quickly or otherwise the nodes step along the curve. To me they represent the pulsation between the infinite and the local, for it reflects the breathing between the immediately accessible and the furthest distances. In Figure 13.14 the curves upon which the nodes lie are all exactly the same.

Nature seems to utilise all these options, so if we wish to analyse a natural artefact it is helpful to use a computer. This means defining the curves in an algebraic form to make programming possible. This greatly speeds up analysis, for in the early 1970s Lawrence Edwards would still calculate with slide rule or log tables. In 1982, having been given a computer, he sent me a small computer plot of the archetypal bud form as developed in this chapter, Figure 13.15. This spurred me to try something similar. All I could then afford was a hand-held Casio programmable calculator. The appearance was poor but the curves were fine.

## 13.3 Plant morphology

The challenge now was to test whether these elegant geometric forms coincide with the visible physical phenomena, not just in profile as we examined earlier in Chapter 10, but in three-dimensional space. Edwards has already done a lifetime of work on this, so my exploration has been partly to confirm some of this for myself, and then perhaps to take it a little further.

On a walk in the bush I discovered the drumstick flower head, an *Isopogon* (Figure 13.17). This little flower head looked like a convex path curve profile, so I began by counting the spiral curves in the two directions. There were 21 spirals clockwise and 13 anticlockwise (Figure 13.18). These of course are consecutive Fibonacci numbers. This is very common in the plant world and has been extensively described (Church, Colman, Cook, etc.), though not really explained. Then, running a path curve program, the spirals were plotted and then matched to the actual flower head. For a first assessment the correspondence, while not perfect, was reasonably satisfactory both for outer profile and for the spirals.

With a protea bud I wondered if I could I place a path curve through the petal tips on *both* spirals that would also give the *same* profile. So I wrote a simple program that could take two points on any curve between the end points of the bud and describe a path curve through these four points. These four points would totally determine such a curve. Plotting an anticlockwise curve (red)

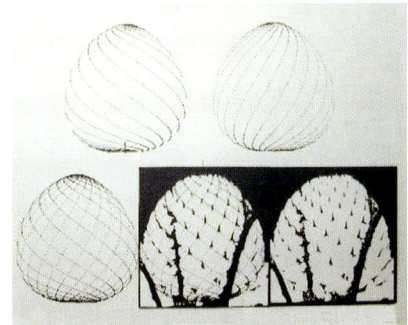

*Figure 13.17 Drumstick flower head, an* Isopogon
*Figure 13.18* Isopogon *analysis showing the two spirals (top), the combined spirals and the actual flower head (below)*

*Figure 13.19 Protea bud*
*Figure 13.20 Analysis of the protea bud* e *is the ε-value and* l *is the λ-value*
*Figure 13.21 General plot where* λ *is 6.0*

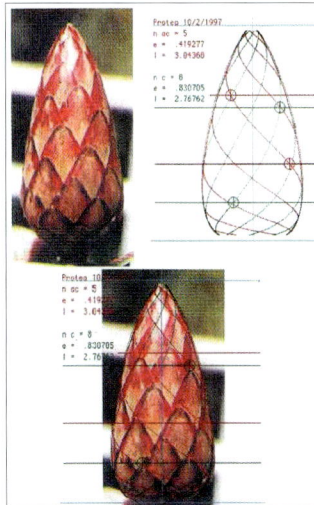

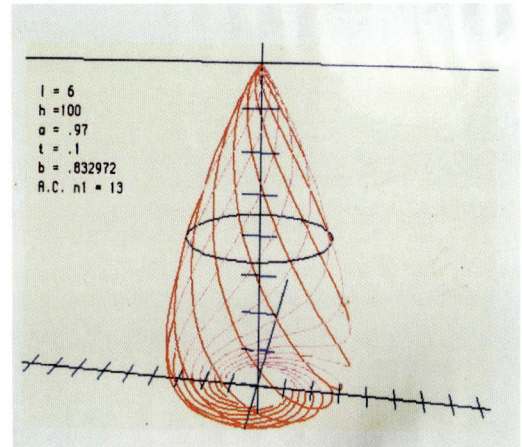

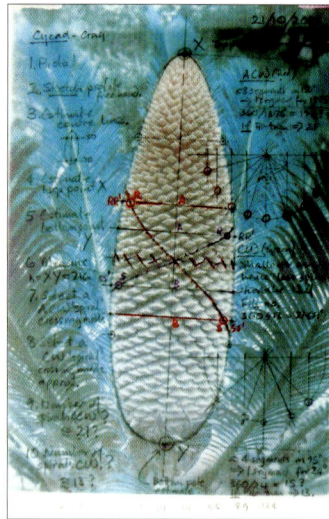

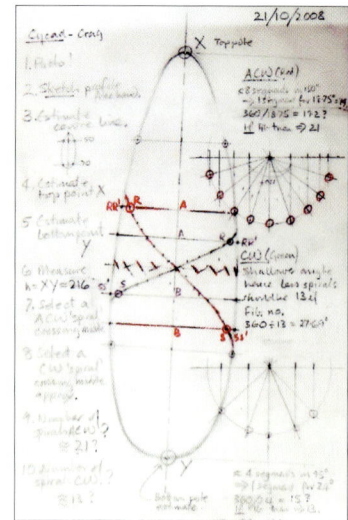

*Figure 13.22 A cycad cone*
*Figure 13.23 Initial analysis on tracing paper*
*Figure 13.24 Process*

*Figure 13.25 Data sheet for anticlockwise spiral*
*Figure 13.26 Data sheet for clockwise spiral*
*Figure 13.27 Cycad and path curve fit*

and, separately, a clockwise curve (green) gave two profiles (Figure 13.20). What astonished me was how close both these outlines were in the overlay: there is barely a millimetre between them. This is an old copy of the original work and is somewhat wanting in clarity.

Obviously, for this to be statistically convincing a number of such buds need to be analysed, but I had neither time nor inclination to do this at that time.

We have many cycads in Australia and the cones looked tempting to try to assess. Going through the process I used for analysis, I first took a photograph (Figure 13.22), sketched around an estimated profile in pencil on tracing paper (Figure 13.23), and then followed

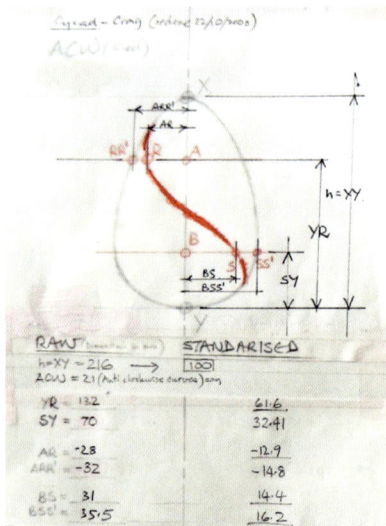

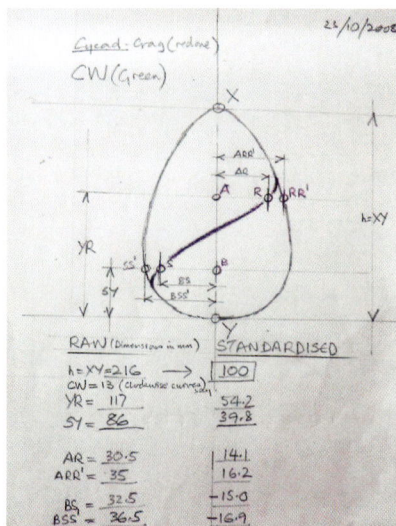

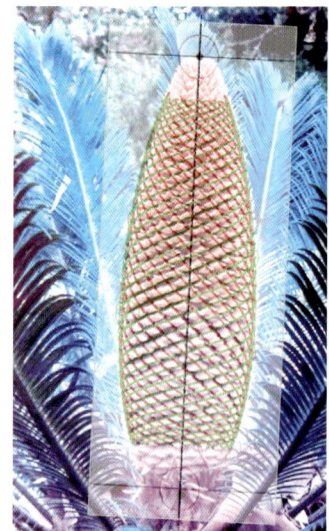

through the analysis by taking two points on a selected anticlockwise spiral (red) and two points on a clockwise spiral (green). Both of these spiral choices were about the middle of the cone (Figure 13.24). The data was recorded (Figures 13.25 and 13.26). Now I compared the plot derived from the program with the actual cone (Figure 13.27). The overlay shows a good correspondence in the middle zone. Generally I only plot about four-fifths of the total height as towards the end points the geometry goes into infinitudes which nature does not and cannot follow. And a close up view of the middle spirals shows that not all seed points coincide precisely but there is overall a good correspondence (Figure 13.28). We would not expect invariable precision, for the plant is growing in an environment with all its vagaries.

*Figure 13.28 Close-up of middle area*

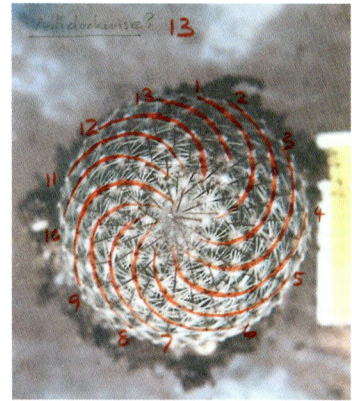

*Figure 13.29* Mammillaria *cactus*
*Figure 13.30 Clockwise spirals*
*Figure 13.31 Anticlockwise spirals*

Another attempt I made at this kind of analysis was of the little cactus, *Mammillaria* (Figure 13.29). Using the view from the top I sketched the spirals in the two directions (Figures 13.30 and 13.31). The anticlockwise curves appeared to be 13 turns and clockwise to be 21 turns. This gave some of the essential input to my program. The remainder came from measurements from the photograph. The result is in Figure 13.32. The curves fitted well through the spiky nodes for slope on one curve. But despite a plausible number count these did not match the actual nodes on other curves. I remain baffled by this, but report it anyway. However the independently determined profiles (only the assumed positions of $X$ and $Y$ are common) did not differ by much, as can be seen from the red and green profiles in Figure 13.33.

*Figure 13.32* Mammillaria *analysis*
*Figure 13.33* Mammillaria *profiles*

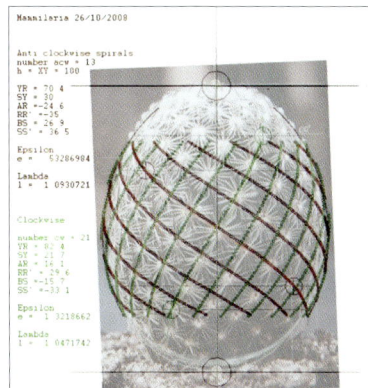

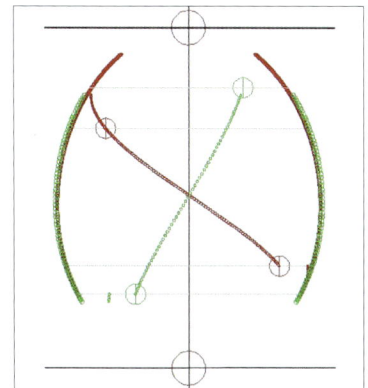

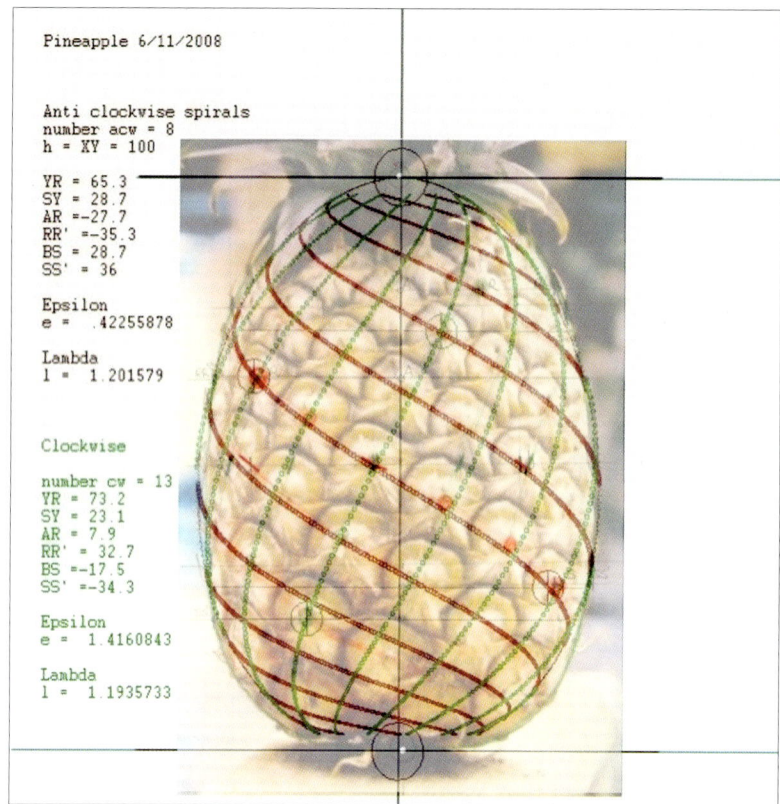

Figure 13.34 Pineapple quadrilateral choice
Figure 13.35 Alternative choice of quadrilateral
Figure 13.36 Pineapple overlay showing a λ-value of 1.20 and 1.19, and λ-value of 0.42 and 1.42

Figure 13.37 Pineapple fish, Cleidopus gloria [Rudie Kuiter]

A first step in such analysis is to decide which spirals to use. There are a number possible. I usually choose the set such that the quadrilateral joining four nearest nodes is the smallest and nearest to square. My first choice in the case of the pineapple is shown in Figure 13.34 with an alternative in Figure 13.35. Following the usual procedure the result is shown in Figure 13.36. At present I only fit two points on any curve. (Someone needs to write a program that will find the best fit for a number of such points.) Again it is the middle two thirds of the form, both in profiles and in path curves through the seed centres, that best meet the actual pineapple. Correspondence towards both upper and lower poles fades away. In the middle zone we see the rhythm of the nodes along the curves, and the number of curves around the centre. These are again two consecutive Fibonacci numbers: 8 (anticlockwise) and 13 (clockwise).

As a little aside, there is a spiky Australian fish called the pineapple fish, *Cleidopus gloria*. I contacted Rudie Kuiter, the author and photographer of *Guide to Sea Fishes of Australia,* who kindly sent me some of his excellent photographs. Just for fun, I wondered how this little fish might respond to this analysis. One has to assume that the cross-section of the fish body is circular, and that

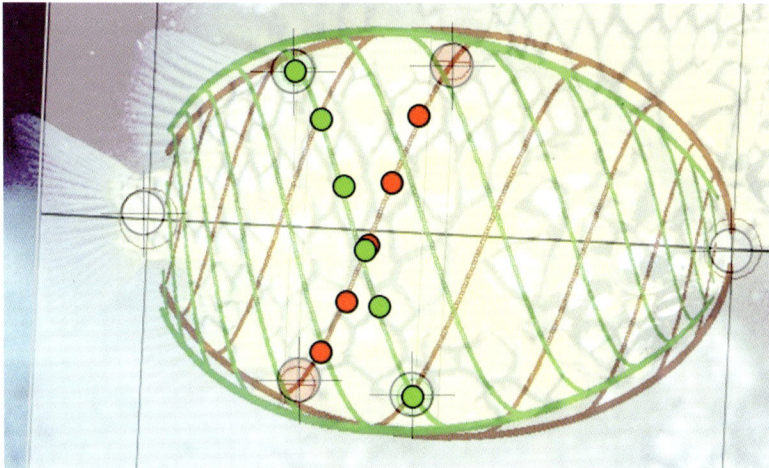

*Figure 13.38 Pineapple fish overlay*

*Figure 13.39 Euglena analysis*
[Mathematics Physics
Correspondence, *No. 19, Easter 1977*]

is of course not the case with most fish. We need to define endpoints, which is difficult. After going through the usual procedures the result showed the profiles are obviously well out, and equally obviously the spiral curves only vaguely matched the centres on very few curves (Figure 13.38). The scales themselves were a different matter again. This was entirely to be expected, but well worth the exercise.

To turn to some tiny examples, Stephen Eberhart, editor of the *Mathematics Physics Correspondence,* analysed the spirals of the microscopic unicellular plant/animal *Euglena*. He did find a λ-value of 1.75 with a deviation of 15%.

There is a tiny little fruiting body on a plant that lives mainly in brackish water, gyrogonite (*Lychnothamunus barbatus*). Peter Glasby who was working on these for his thesis brought them to my attention. They almost certainly show a path curve, though the curves only go anticlockwise. (Before the Devonian period they spiralled in the opposite direction.) They are only about 1 mm (¹⁄₂₅ in) long but have a well-defined structure. Again we find that in the central two thirds of their length there is a good match. The plotted curves (intended to be the spiral curves at the bottom of the ridges) coincide quite well with the real thing.

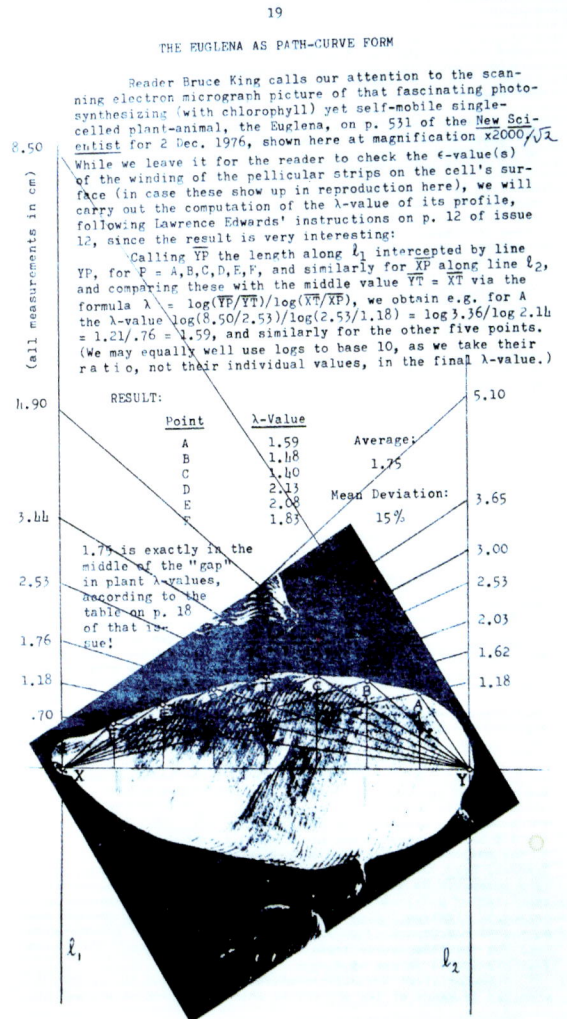

19

THE EUGLENA AS PATH-CURVE FORM

Reader Bruce King calls our attention to the scanning electron micrograph picture of that fascinating photosynthesizing (with chlorophyll) yet self-mobile single-celled plant-animal, the Euglena, on p. 531 of the New Scientist for 2 Dec. 1976, shown here at magnification x2000/√2. While we leave it for the reader to check the ε-value(s) of the winding of the pellicular strips on the cell's surface (in case these show up in reproduction here), we will carry out the computation of the λ-value of its profile, following Lawrence Edwards' instructions on p. 12 of issue 12, since the result is very interesting:

Calling $\overline{YP}$ the length along $l_1$ intercepted by line YP, for P = A,B,C,D,E,F, and similarly for $\overline{XP}$ along line $l_2$, and comparing these with the middle value $\overline{YT} = \overline{XT}$ via the formula $\lambda = \log(\overline{YP}/\overline{YT})/\log(\overline{XT}/\overline{XP})$, we obtain e.g. for A the λ-value $\log(8.50/2.53)/\log(2.53/1.18) = \log 3.36/\log 2.14 = 1.21/.76 = 1.59$, and similarly for the other five points. (We may equally well use logs to base 10, as we take their ratio, not their individual values, in the final λ-value.)

RESULT:

| Point | λ-Value |
|---|---|
| A | 1.59 |
| B | 1.48 |
| C | 1.40 |
| D | 2.13 |
| E | 2.08 |
| F | 1.83 |

Average: 1.75

Mean Deviation: 15%

1.75 is exactly in the middle of the "gap" in plant λ-values, according to the table on p. 18 of that issue!

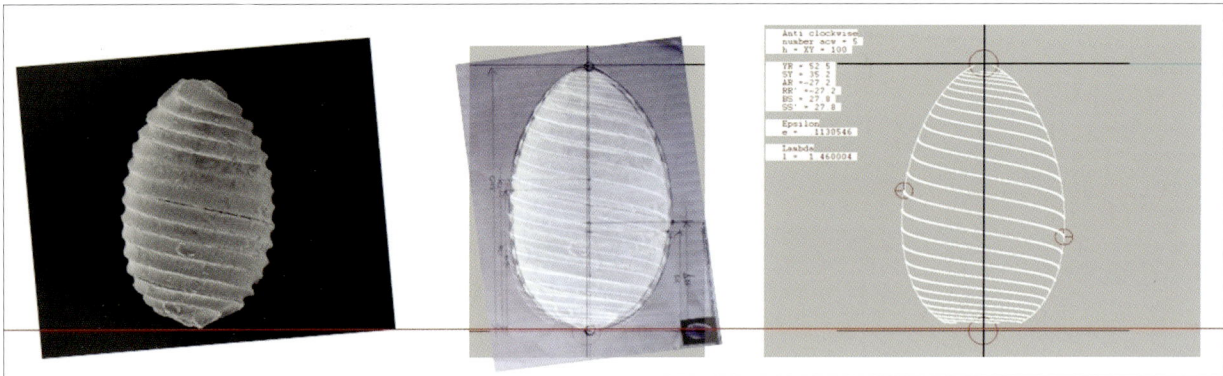

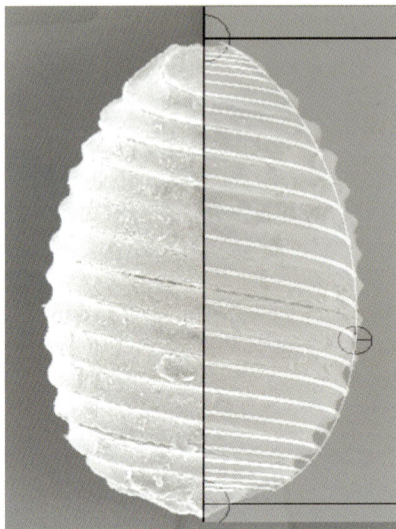

Figure 13.40 Gyrogonite
(Lychnothamunus barbatus)
Figure 13.41 Split image comparison
is clearer: the left side is the actual
gyrogonite, the right side shows the
theoretical spirals [base image: Adriana
Garcia, University of Wollongong]

The forms we have described until now have the sharper end at the top, the blunter at the bottom. However, there are some instances of the opposite, particularly in many flower forms at a stage in the opening process. We shall look at the petal tips of the protea flower. Once again a consecutive pair of Fibonacci numbers is assumed to fit the case, in this instance 8 spirals anticlockwise (steeper slope) and 5 clockwise (Figure 13.43). As the form is the other way up, the $\lambda$-value, at 0.61, is now less than 1. Again the fit is not so good near the two ends but is quite close to the actual petal tips (marked as small black dots) in the middle region.

In the next example we have a flower that is open at the top, and only the lower part shows the bud shape. To analyse this we have to imagine an uppermost point in the physically empty space above. This unidentified flower is from Western Australia. Many years ago I had estimated a profile and plotted it (Figure 13.45), but at that time I had not worked out a way to examine the placements of the petal curves. Now, on analysing the spirals, we see that, even though the curves do not go exactly through the marked points, they lie on virtually the same slopes, both anticlockwise (red) and clockwise (green).

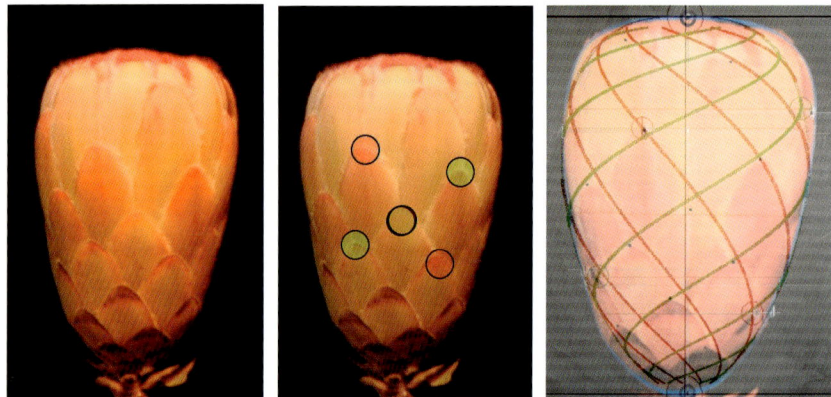

Figure 13.42 A protea flower
Figure 13.43 Choosing key points of
the spirals
Figure 13.44 Protea spirals

## 13.4 Fields of morphology

It is astonishing that so much of nature follows this geometry, as if this meeting of space and counterspace,* the real and the imaginary, produces a morphological field fundamental to the vegetative and plant world. The fields, that may be akin to Rupert Sheldrake's morphic fields, are theoretically of huge extent and interacting with countless other such fields. Something of that is shown in Figure 13.47 where the 'bud' in the centre is only a small part of the whole construction. A single curve is shown, but any number of others of the family could be drawn in the same field. And the two curves above and below the parallel lines are actually one and the same curve going through infinity. (Only our imagination has difficulty with this, the geometry is simple.)

* See the work of Nick Thomas, Louis Locher-Ernst, George Adams.

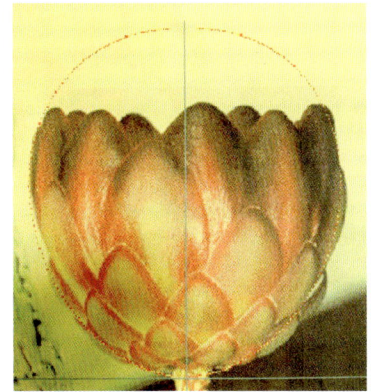

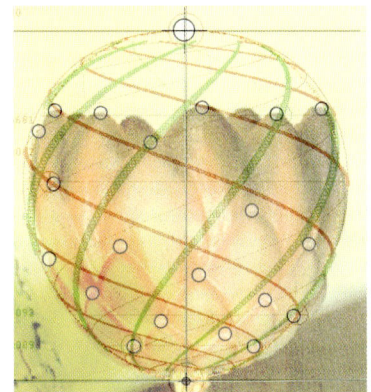

Figure 13.45 An opening flower
Figure 13.46 Overlay of the curves

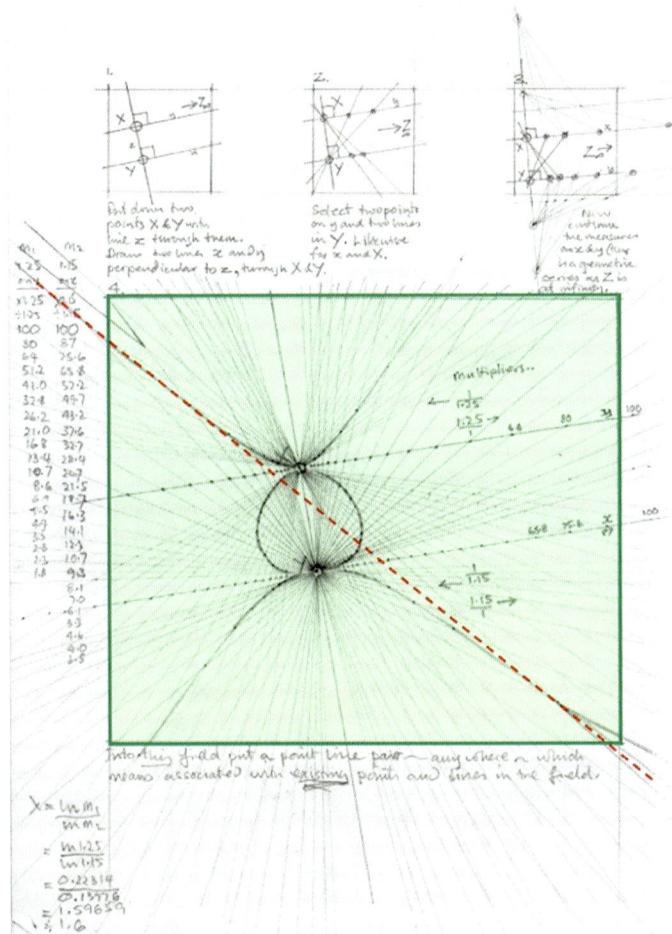

Figure 13.47 Bud profile construction
Figure 13.48 Sketch of single path curve surface

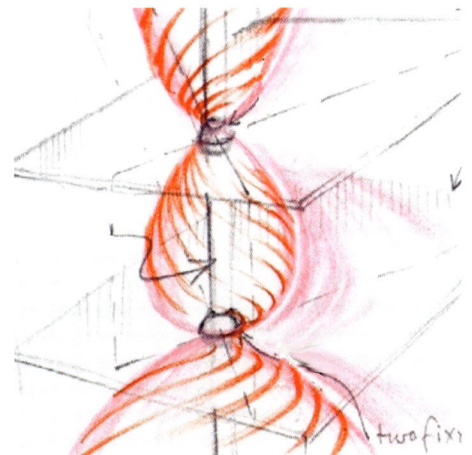

Figure 13.49 An opening bud sequence

In space (not just in a plane) the field is almost impossible to draw – I have sketched just *one* surface of such a spatial field in Figure 13.48.

But needless to say, as vegetation grows it changes. As a bud opens, more and more of the total form becomes invisible, even though the visible part becomes larger (Figure 13.49). Recent and current work suggests that this change can be followed, and that the form factors change: the $\lambda$-value and perhaps also the $\varepsilon$-value. A tentative exploration of this changing form follows.

## 13.5 Bud transformations in time

We are all familiar with flowers opening and closing every day, even if it goes largely unremarked. In this simple opening and closing is obvious evidence of change in the plant forms. Can this kind of change be understood a little more as an expression of definite transformations?

Lawrence Edwards watched these changes with oak and beech buds, and many others. He set out one day to check some previous work and found that there was a slight change in the $\lambda$-value of his results. He discovered that the oak buds had a slight pulsation of form during the apparently dormant time of about six months until the leaves suddenly burst out. The buds gradually got a little larger for the first few months, then settled down to a virtually constant size, but not a constant form. A pulsation was revealed in the small changes of the $\lambda$-value over these months. There was a temporary and small reduction in the value every fourteen days or so. This means that the bud profile tends slightly towards an elliptical shape or profile, that is, the sharpness at the top pole decreases a little. So at these intervals we get a graphical picture that tends to show dips at relatively regular times. He wrote this up in his *The Vortex of Life* and in a number of supplements.

The oak buds appeared to pulsate to a fortnightly rhythm. There would be dips in the $\lambda$-value about every two weeks. At the time Lawrence discovered this, the rhythm correlated with

Opposite:
Figure 13.50 Oak tree in Chaleyer Street, Sydney
Figure 13.51 Oak bud
Figure 13.52 Oak bud about 6.5 mm (¼ in) high
Figure 13.53 Pentax camera with Tamron macro lens and mounting
Figure 13.54 Analysis of an oak bud

the times when the moon was in conjunction and opposition (that is about every two weeks) with the planet Mars. Every little oak bud was doing a little fourteen-day dance in tune with the moon and Mars. At first sight this seems surprising, but when we consider that the geometric field surrounding each bud stretches to infinity it is perhaps not implausible that it encompasses the sphere of Mars. However, over the years there was a discernible shift, leading to a much more complex situation than a simple one-bud-to-one planet correlation. Graham Calderwood elaborates on this in the 2006 edition of *The Vortex of Life.*

In Australia the oak, *Quercus robur*, is a feral plant having been imported from Europe. However, I have been photographing the buds of a large nearby tree. The position of this tree is 33.35°S latitude and 151.1°E longitude; this may or may not be important but is noted in case the phenomena are related to latitude or even longitude.

To photograph the buds I removed a few leaves very close to the bud, being careful not to interfere too much with the strength or vitality of the tree. The buds are small. From January to August (through the southern summer and autumn) they change in length from about 5 mm to over 7 mm (⅕ to ¼ in). The ones chosen are usually those sitting on the ends of a low, accessible branch.

I select a minimum of three buds each year. These are photographed daily whatever the weather. The time (3.30 pm), the date (February 19, 2003) are noted. I usually took the picture between about noon and 4.30 pm. I believe that the actual hour is not too significant, but that still remains to be tested. The camera is set at f/8 aperture and 1/60 seconds shutter speed. The bud is inserted in a small hole about 150 mm (6 in) in front of the lens (arrow in Figure 13.53). I still use 100 ASA Fuji colour film. Black and white would be adequate for the profiles of the buds, but perhaps some day other aspects may be revealed, and you cannot retrospectively manufacture data. For the moment we are only interested in the profiles of the buds.

Once developed, I send the pictures to Graham Calderwood in Scotland for analysis. Graham has developed some excellent software to analyse bud (and other related) profiles.* The program can give the λ-value for the oak bud after the cursor traces around the profile, and the degree of error is also given. After examining the profiles for each day and after many months the changes in the λ-value can be set out in a graph. The λ-value does not change a great deal but there is a perceptible rhythmic change.

* Graham Calderwood's program is called Bud Workshop and is available from his website *budworkshop.co.uk*

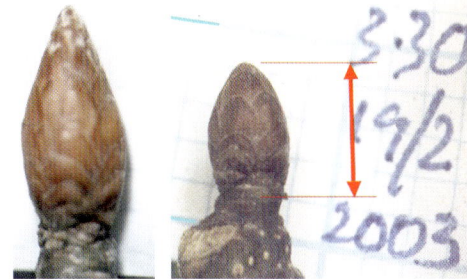

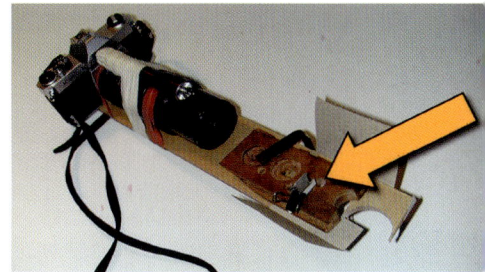

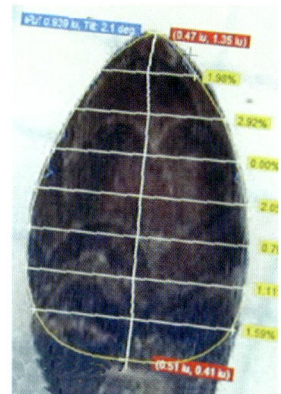

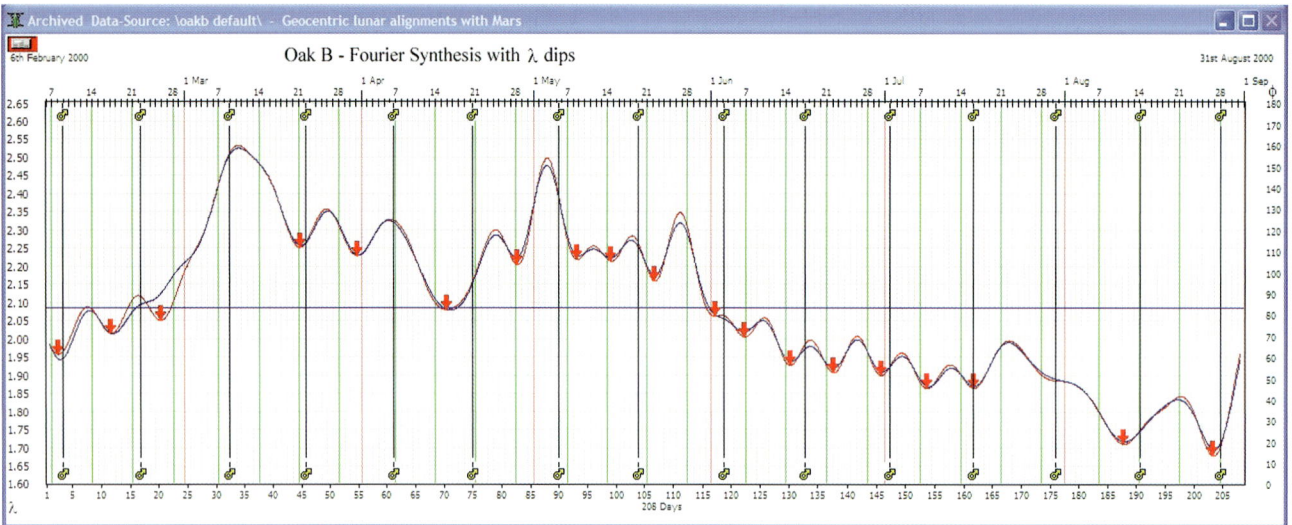

*Figure 13.55 Analysis of the λ-value of an oak bud*

*Figure 13.56 Variation in the mean λ-value of beech buds observed in Strontian, Scotland, 1991. The alignments of Saturn with the moon are shown as little down-turned triangles [Edwards,* The Vortex of Life]

The graph (Figure 13.55) shows a regular change, but unfortunately the oak bud did not show an allegiance to Mars! For some time it seemed to relate to rhythms of Saturn or Jupiter and then switched to Venus. Obviously much more research needs to be done before any conclusions can be reached.

Edwards's work over at least fifteen years suggested this correlation not just for oaks but for other plant buds and planets too.

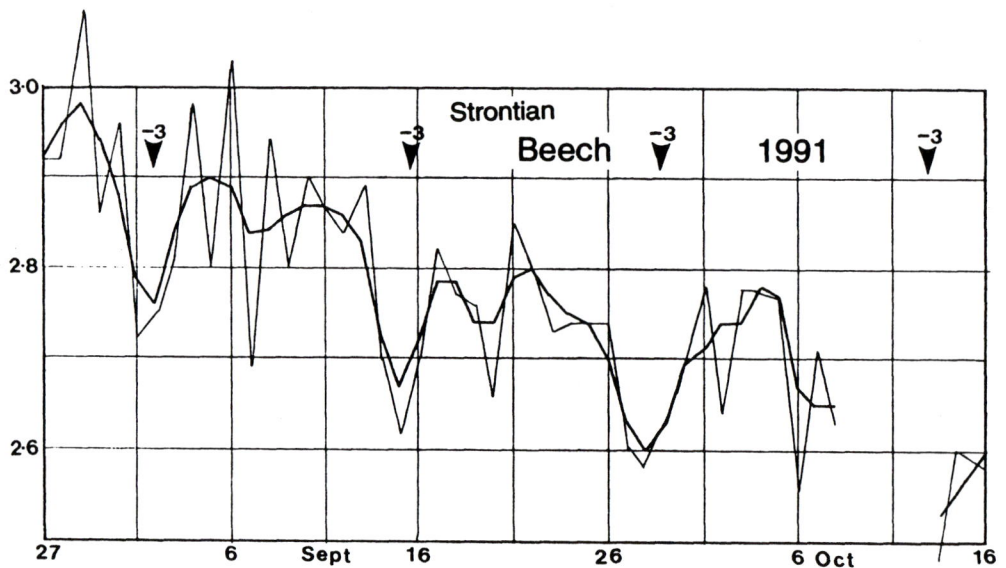

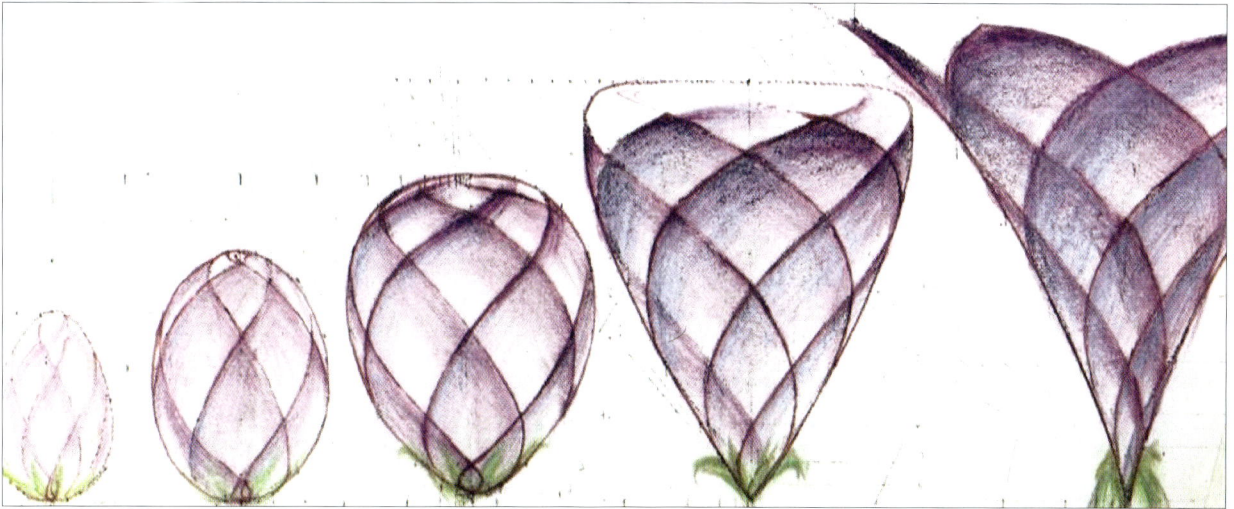

Figure 13.57 A speculative sketch of the sequence of cycad frond opening

He found that the beech tree appeared to respond to the rhythm of moon and Saturn, the cherry tree to the sun. An example from his work is shown here for the beech tree (Figure 13.56).

## 13.6 Cycad frond transformation

Over many years I have observed the striking and beautiful cycads (which we looked at earlier in section 13.3). When the fronds begin to appear they show as a little dome amidst the previous season's chalice of older fronds. This dome develops

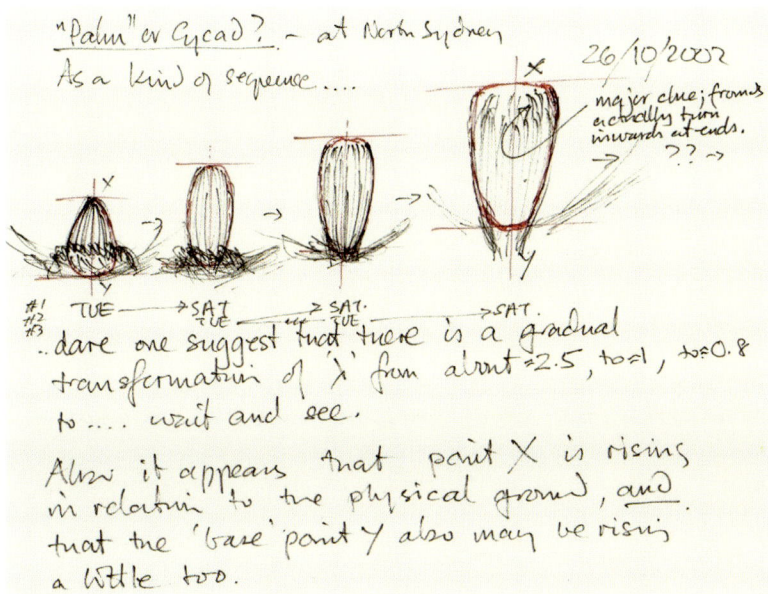

Figure 13.58 Notes of a sequence

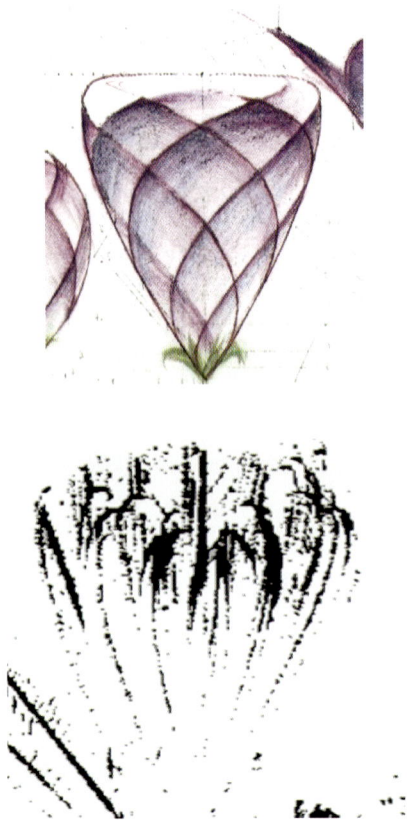

*Figure 13.59 The crucial nearly conical stage*
*Figure 13.60 The cyclad frond at the crucial moment*

*Opposite and overleaf:*
*Cyclad fronds opening up, with sketch of bud forms every two days*

rapidly into an upright elongated egg shape, and then starts to become more bulbous at the top until it becomes almost conical. The fronds then start to bend away from the centre line and the whole thing begins to open up becoming vortex-like. I could imagine the transformation in sequence (Figure 13.57).

I made some weekly observations (Figure 13.58). What was really crucial was whether this plant went through all these imagined steps. I eagerly watched it, especially for the stage where the form suggested an inverted bud but was also open at the top. If the sequence was plausible there had to be a stage where the frond tips would be slightly turned inwards but the space open in the middle. This would be just before the frond became conical, with a λ-value of 0, before it opened up and the λ-value becoming negative.

Would it do this? If it did it would be a clear indication that the sequence was like the one imagined. Well, it most certainly did do this, and I have since then observed this many times (Figure 13.60). This was a 'eureka' moment! Notice that the tips of the fronds are curling inwards much as predicted, yet the centre is open and empty. After this the form gradually turned into a vortex-like structure.

Now I needed to get much more data on all this. Despite the practical difficulties I managed a few sequences. A selection of these images is shown consecutively in Figures 13.61 to 13.78. Obviously the correspondence is not exact but there is an undeniable trend there. Once the fronds really opened up into the vortex form the form followed the geometrical vortex a bit more closely.

A graph plotting the λ-value against time showed a fairly straight-line decrease that seems a little simple for nature's usual more complex working. I more or less caught the visible start of the process (on November 11, 2005), and after December 3 – a mere 22 days, the plant had begun its vortical fan and increased hugely in size. I also observed the fine leaves on the major frond stems. Even on December 7 these had not fully straightened and were still quite soft and fragile, showing there were still some changes going on.

I had expected a slow and gradual opening up of the fronds over several days or weeks, and I thought that this gradual process would even be noticeable during the course of a single day. So I took pictures every few hours expecting a gradual change. I was quite surprised to find they opened up slowly and then gradually closed a bit. Here was a discernible breathing or pulsation I had not looked for.

13.61 Nov 8, 2005

13.62 Nov 9, 2005

13.63 Nov 10, 2005

13.64 Nov 11, 2005

13.65 Nov 12, 2005

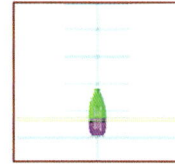

Nov 11

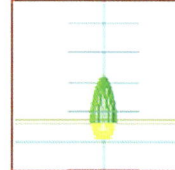

Nov 13

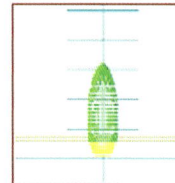

Nov 15

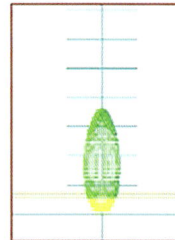

13.66 Nov 13, 2005

13.67 Nov 14, 2005

Nov 17

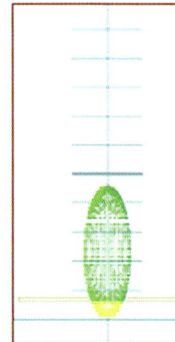

13.68 Nov 15, 2005

13.69 Nov 16, 2005

Nov 19

13.70 Nov 20, 2005

13.71 Nov 21, 2005

13.72 Nov 22, 2005

Nov 21

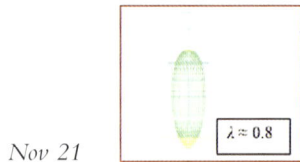

$\lambda \approx 0.8$

Nov 23

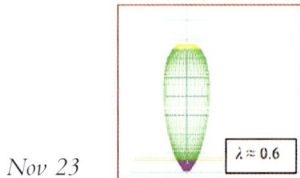

$\lambda \approx 0.6$

Nov 25

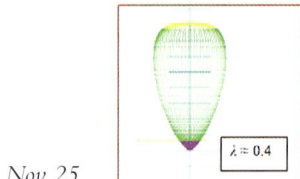

$\lambda \approx 0.4$

Nov 27

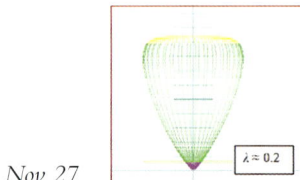

$\lambda \approx 0.2$

Nov 29

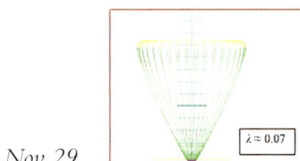

$\lambda \approx 0.07$

Dec 1

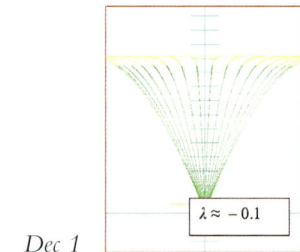

$\lambda \approx -0.1$

13.73 Nov 23, 2005

13.74 Nov 24, 2005

13.75 Nov 25, 2005

13.76 Nov 26, 2005

13.77 Nov 27, 2005

13.78 Nov 28, 2005

Much more work needs to be done for these observations and tentative conclusions to be credible. But they throw open further questions. For instance, with the cycad, do the upper and lower poles of the field shift, and if so, how? One could have the impression that the invisible form as a whole rises through the plant, leaving a trail behind it of less affected older frond portions. Broader questions abound too. What kind of forces draw this plant forth? Or, from another point of view, what being is expressing itself through the visible form? In *Space and Counterspace* Nick Thomas refers to stresses between two worlds of force – between our everyday space and counterspace through which formative forces are active. Is the physical growth of the plant a resolution of this stress?

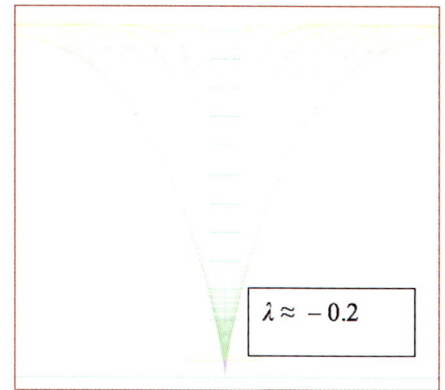

$\lambda \approx -0.2$

*Figure 13.79 Dec 3*

# 14. Form in the Animal World

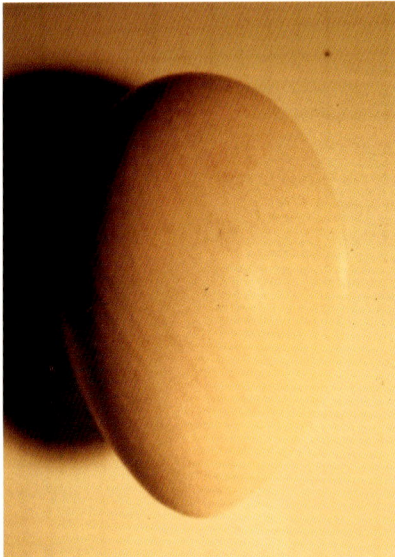

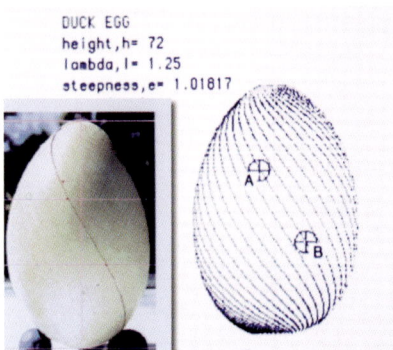

The geometric forms we have looked at in the mineral and plant world are generated by growth measures along the edges of a tetrahedron. In the case of the mineral world it was an infinitely large, all-real tetrahedron, and for the plant world it was a complex, or semi-imaginary one. There is the question of whether there is a kind of tetrahedron that generates the forms we find in the animal world.

Some of the organs of animals, like the heart, respond well to the tetrahedral structure as work by Edwards and Calderwood shows. Generally though we can see that in the step from the mineral to the plant world, translation symmetry is lost. The plant world shows rotation symmetry, but much of that is lost in the step to the animal world. Here the emphasis seems to be on the bilateral, the mirroring, symmetry.

As we saw earlier the orientation of the animal world is predominantly horizontal. There is the nodal (or point-like) emphasis at neck (larynx) and hips (reproductive area). There are vertebrae, the rhythm of the nodes between these two. Along this spine there is a transformation in every one of these consecutive vertebra.

If there is a tetrahedron for this realm, there must be a visible horizontal line, and a second line (perhaps perpendicular to it). For the mineral the six lines of the tetrahedron are all in the infinite sphere. For the plant, one visible line vertical to the earth's surface has its counterpart in the perpendicular line, the second line, horizontally in the infinite. The other four connect or weave between these two extremes and are imaginary, in circling motion. From mineral to plant just one line has become local, the other remains in the infinite. From plant to animal, there has been a further change, I believe. But where are the other four lines? I am speculating of course, and I have to admit the animal tetrahedron still eludes me.

DUCK EGG
height, h= 72
lambda, l= 1.25
steepness, e= 1.01817

*Figure 14.1 Duck egg with a hint of spiralling*
*Figure 14.2 Duck's egg path curves*

## 14.1 The spirals of the egg

We have seen the bud-like form in the animal world when we looked at eggs (section 10.5). The egg seems to be *the* form of archetypal beginnings and is an early stage in the life cycle of many animals. We saw there the hint of a spiral on one of the eggs. It can also sometimes be seen on duck's eggs. The spirals must have something to do with the duck's oviduct, the path through which the egg is laid. They are akin to the two sets of spirals on bud surfaces, but on the egg there is only one set, an anticlockwise one. Early work established a $\lambda$-value of 1.25 and an $\varepsilon$-value of nearly 1. This latter appeared to be about right as the slope was only a bit more than 45°. A singular path curve has been plotted in the picture on the left in Figure 14.2 and a set of curves has been computer generated on the right to give an overall impression.

In *The Vortex of Life* Edwards notes wryly, 'It will be seen that the humble hen ... producing forms with an MRD [Mean Radius Deviation] of one half of one percent, was working three time more precisely than I was able to do with all those years of experience behind me. The actual deviation on the egg itself lies in the neighbourhood of one hundredth part of an inch – at the very limits of measurement even using a lens.' (p. 64)

The Port Jackson shark's egg is often found on the New South Wales coast. Its basic outline is a definite egg profile, but it has fins that spiral round the outside and have counterpart curves on the inside. The direction of spiral is clockwise. Why do ducks and sharks have a different spiral direction in their eggs?

Douglas Baker in the United States has checked 250 different species of bird eggs for a path curve agreement and with much good agreement too. The ostrich egg shown has faint spiral marks (Figure 14.5) as we saw earlier (Figure 10.29). Usually these spiral lines are not so visible on bird's eggs, though the profile itself strongly points to these path fields being at work.

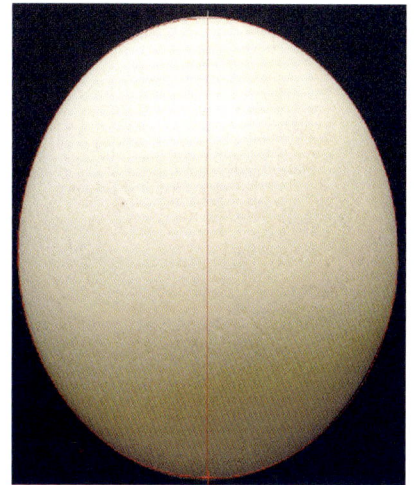

*Figure 14.4 Inside the shark egg (fins removed)*
*Figure 14.3 Port Jackson shark egg*
*Figure 14.5 Ostrich egg with faint spiral marks*

Figure 14.6 Forktailed large-eye bream

## 14.2 Fish

The fish has a predominant horizontal line. Can we still call upon path curves to help us grasp something of the animal form? We could almost be excused for seeing the fish as a slim horizontal pine cone form with jaws at one end and a tail at the other, and a few fins poking out in between. Both broadly fit into the same oval profile, but are 90° apart in their natural orientation.

To begin with, let us look at a bream. Superimposing an elliptical outline on the form we find it is not too bad a fit, and might indicate a path curve. The sleek unicorn fish is blunter at the head and sharper at the tail end. The overlay here is of a path curve with a λ-value of about 2. This gets a little bit closer to the more general fish profile, but is still far from satisfactory.

It would appear that there is going to be a difficulty finding a match for these fish forms with the basic path-curve profile so far explored. And, of course, a fish is not a plant. There is the further question of symmetry. The egg or bud path curves expressed in space have a circular or radial symmetry. The generating curves for the bud or egg all rotate about the central vertical axis, the stem. Hence the curves appear to spiral about this central line of the plant spine. The end-on view of a fish, however, is not circular, but it does have bilateral or reflection symmetry about a vertical plane running through head and tail.

*Opposite, from top:*
*Figure 14.9 Spangled emperor showing the curved lateral line distinctly [Rudie Kuiter]*
*Figure 14.10 Eastern nannygai showing an almost straight lateral line*
*Figure 14.11 Section through a salmon (coloured for emphasis)*
*Figure 14.12 Trevally*
*Figure 14.13 Sword fish*

*Figure 14.7 Sleek unicorn fish [Rudie Kuiter]*
*Figure 14.8 Overlay of a path curve on the profile*

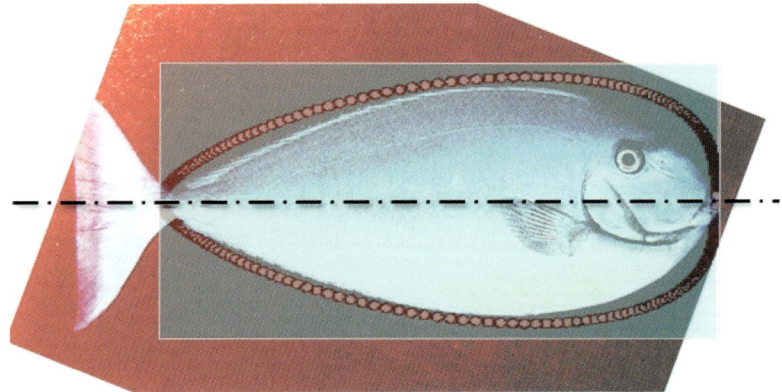

The tetrahedral architecture that might provide the basis for the form of fish must be quite different to that of the plant, as the tracery of curves does not circle round the spine but appears to meet in two ridges at the top and bottom, dorsal and ventral, of the creature's body. There is also a seam or suture along the middle upper body of many fish. Is this the lateral line? Is this a hint of the overarching architecture? This lateral line is well known and appears to be connected to the method by which the fish orients itself. It tends to be along the top two thirds of the fish body and curve a little downwards at the head end of the creature. In some species this downturn is very marked, in others it is almost straight. This line or curve may be an inherent part of the basic geometric structure.

For the mineral forms all six lines of the tetrahedron are at infinity. For the plant one line remains in the infinite, another becomes local (the stem or trunk), and the other four lines are imaginary, or in movement. For the animal world, I have wondered whether the line that is at infinity in the plant, becomes progressively more local in the animal. Something more is working in the animal realm than the mere mineral laws or even the laws of vegetative growth. The animal has an emerging consciousness, or *animus*. Does this perhaps work into the forming of the structure and shape of the creature?

Returning to the fish and its lateral line, in cross-section the fish body has definite whorls of flesh and muscle above and below this outer line. A section through a salmon shows that the lateral line is indeed significant. It came at a place where the inner formations making up the body of the creature had a noticeable horizontal partition or septum, and this partition went right through the spine at the section shown. This lateral line divides the body into above and below. Above and below this horizontal partition are whorls in the flesh that are definitely not reflections of each other. There is no symmetry of above and below.

Is it worth testing some hypothetical tetrahedron, with its inherent path curves, for consonance with the actual fish architecture? The architecture needs to allow bilateral curved surfaces, symmetrical about a vertical plane, the two separate surfaces (left and right) being laced with sets of path curves which cross each other in such a way as to be able to give a morphological/geometrical basis for the skin and scale patterning seen on numerous fish.

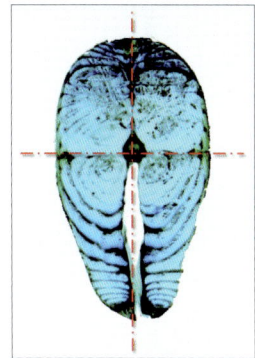

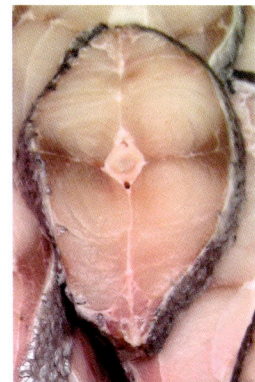

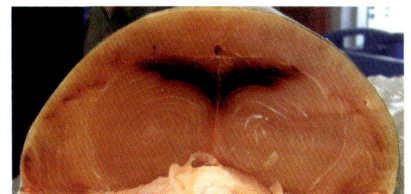

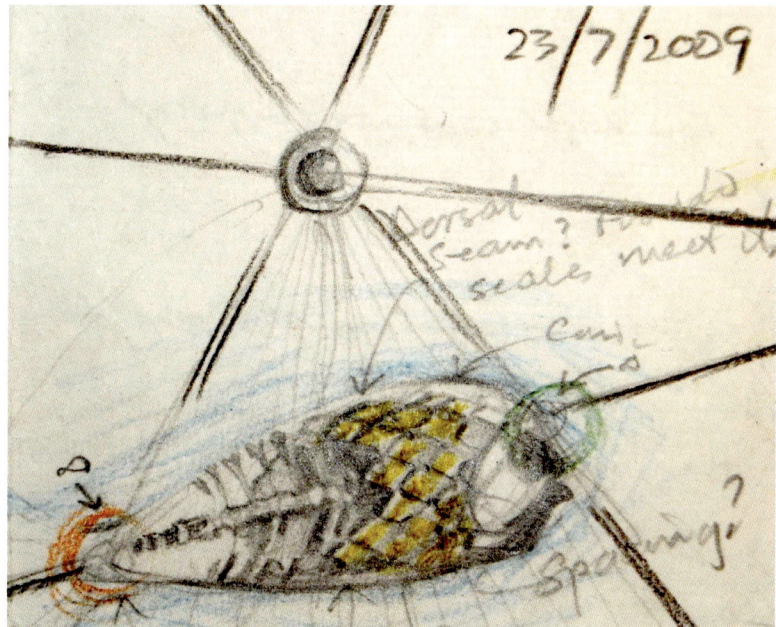

*Figure 14.14 Fish form within a trial tetrahedron*

## 14.3 The tetrahedron for the fish form

I sketched a proposed tetrahedron with a fanciful fish form and even more fanciful scale pattern (Figure 14.14). Can these curves of this initial tetrahedron mimic the scale pattern? The dorsal and ventral seams on the fish body would need some limiting curve as the scales do sometimes get smaller towards the upper and lower seams, but I was still not convinced.

These notions led to the tetrahedron suggested in Figure 14.15. It is basically a tetrahedron which has been flattened into a triangle, by making two points ($P_3/P_4$) coincide as the point above the fish, this will make the opposite two planes ($\pi_1/\pi_2$) coincide, and there will

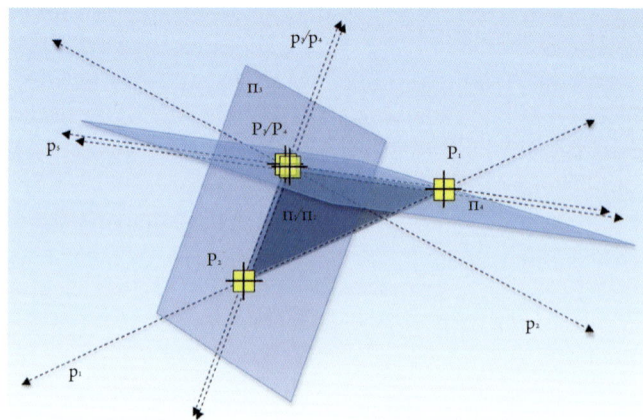

*Figure 14.15 A trial tetrahedron*

also be two pairs of coinciding lines ($p_3/p_4$ and $p_5/p_6$). The triangle has point $P_1$ at the tail, point $P_2$ at the head and the double point $P_3/P_4$ somewhere above. In this tetrahedron the line at infinity (in the plant's tetrahedron) is now local.

In the line of the spine, $p_1$, would be the familiar growth measure, while perpendicular to it in line, $p_2$, could be a step measure. Two coincident planes $\pi_1/\pi_2$ would allow bilateral symmetry, a reflection. Figure 14.16 shows this basic framework, hinting at some of the planes, and showing some of the measures.

There will be path curves in the double plane determined by the triangle $P_1$ (tail end), $P_2$ (head end) and $P_3/P_4$ (double point at the top somewhere). The double lines $p_3/p_4$ and $p_5/p_6$ both go through the double point $P_3/P_4$. The remaining line $p_2$ of the tetrahedron is perpendicular to the vertical double plane and above the line $p_1$. These path curves will be as in Figure 14.17. The yellow dots show a plausible fish profile. With this kind of path curve the form options increase hugely, for the profiles can be more pointed/blunt, as well as there being all sorts of possibilities of asymmetry. Some fish species may have the 'top' line underneath and below the spine.

There are two more planes, $\pi_3$ in front, and $\pi_4$ at the back. Here the path curves are quite different as the measure in line $p_2$, is a step measure, not a growth measure. I have drawn them symmetrically about the vertical plane (Figure 14.18). Only one set of the path curves has been drawn here. This image is like a front view looking along the spine (marked as a coloured circle).

Would such an image correspond with the actual section of a fish? There were certain difficulties in checking this. Few such pictures of front views can be easily found, so I tried to photograph some in the Sydney aquarium. Despite their fast and unpredictable movements I managed to catch some fish in end-on view. Analysis of one showed a $\lambda$-value of 1.7 (Figure 14.20 and panel).

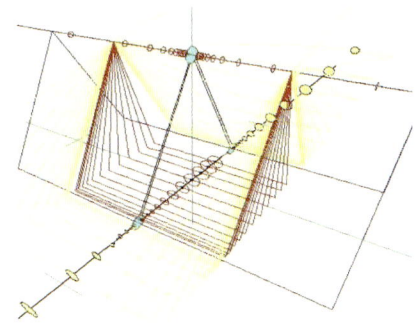

*Figure 14.16 More of the trial tetrahedron*

*Figure 14.17 Planar path curves for an all-real triangle in a vertical plane*

*Figure 14.18 Path curves in front plane, the view along the spine*

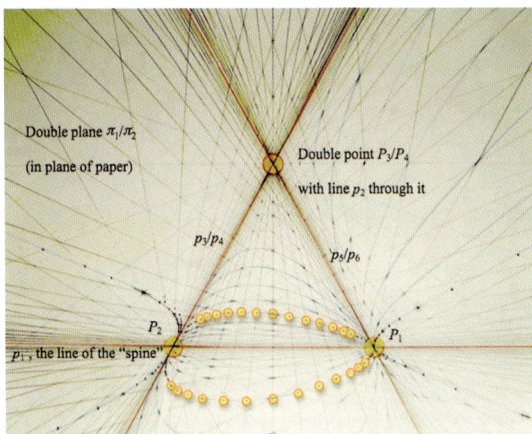

Double plane $\pi_1/\pi_2$
(in plane of paper)

Double point $P_3/P_4$
with line $p_2$ through it

$p_3/p_4$

$p_5/p_6$

$P_2$

$P_1$

$p_1$, the line of the "spine"

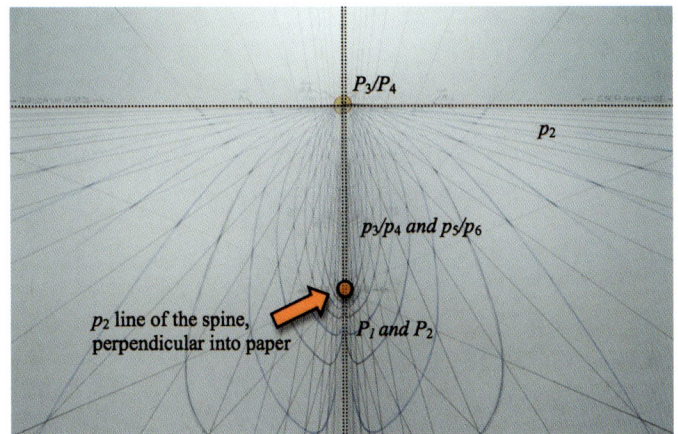

$P_3/P_4$

$p_2$

$p_3/p_4$ and $p_5/p_6$

$p_2$ line of the spine, perpendicular into paper

$P_1$ and $P_2$

1  Draw the outline freehand in pencil on the
   photograph.

2  Estimate and draw the position of the central
   vertical axis by halving the width in a couple of
   places.

3  Where this vertical line crosses the top and bottom
   of the freehand profile mark two points, $X$ and $Y$,
   respectively. These are the dorsal and ventral
   points of the section at its largest.

4  Draw a line through top and bottom points
   perpendicular to the vertical axis. These are the
   parallel horizontals.

5  Now mark two points on each side of the profile,
   about one quarter and three quarters of the way
   along the vertical.

6  From the top and bottom points draw lines
   through these four points as shown, extended to
   cut the two horizontal parallels.

7  From $X$ and $Y$ to these eight cutting points,
   measure the horizontal distances.

8  The multiplier along the right hand side of the top
   line will be
   $m_1 = 199/43 = 4.627$
   and along the bottom will be
   $m_2 = 143/58 = 2.466$.

9  The $\lambda$-value is
   the ratio of the logarithms of these values
   $\lambda = \log m_1 / \log m_2$
   $\lambda = \log 4.627 / \log 2.466$
   $\lambda = 1.697$.

10 For accuracy the left hand side can be similarly
   checked.

*Opposite, top:*
*Figure 14.19 Frontal view of a fish*
*Figure 14.20 Fish profile analysis*

*Opposite, bottom:*
*Figure 14.21 An idealised path curve with*
*λ-value of 1.7*
*Figure 14.22 A good fit between calculated*
*and actual form*

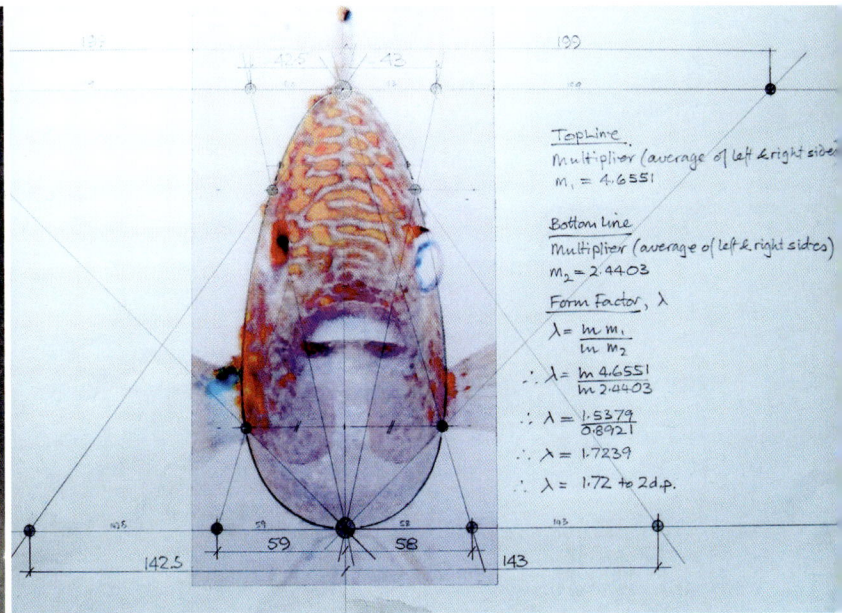

The calculated result needs to be checked against the actual form. So I ran an envelope plot, that is one with tangent lines, with the calculated λ-value (Figure 14.21). Overlaying this on the photograph shows a good fit (Figure 14.22). It was encouraging to find that at least some fish cross-sections corresponded to a path curve approach. Some fish cross-sections show an indentation at the sides, and cannot be egg-like path curves. They may be asymmetrical Cassini ovals that can vary in shape from oblong ovals to lemniscates, but they may be something other.

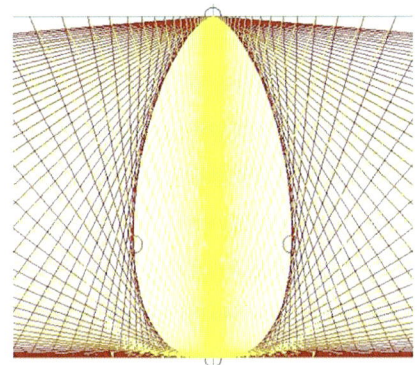

## 14.4 The scale pattern

The scale patterning on some fish is reminiscent of the pine cone. Is there more to it than superficial appearance? We now have a profile which is worth trying in the space fields of a tetrahedron.

I initially decided to build a model of the tetrahedron which was flattened into a triangle. For simplicity I chose an equilateral triangle. The measures (or rhythms) were to be a single step measure in the top line, $p_2$, and growth measures both in the horizontal spine, $p_1$, and in the double lines linking these two. I introduced a section of a fish already measured into the model,

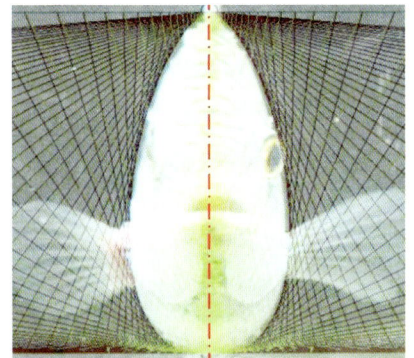

Figure 14.23 Carp scales
Figure 14.24 The abandoned model of the 'flattened' tetrahedron
Figure 14.25 One set of complementary path curves
Figure 14.26 Trial forms giving unlikely curves

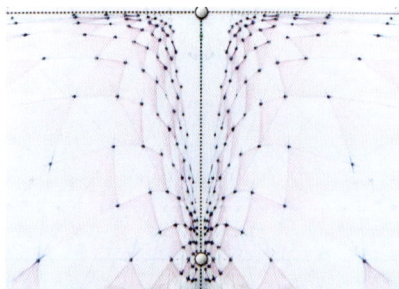

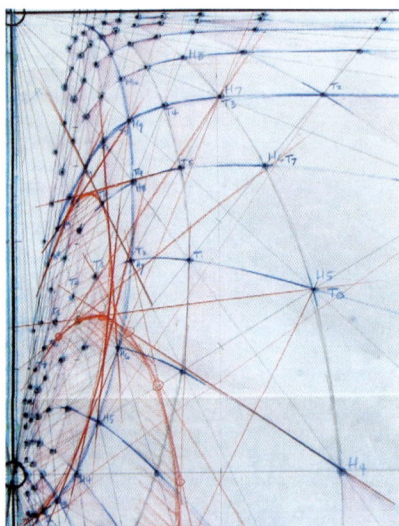

and from that constructed spatial path curves. Would this curve of tangents look remotely like a series of scales?

It is important to choose a suitable point to begin with. I found an arbitrary choice did not get far with useful forms. Despite trying various things I could not get the result to look remotely like a serious fish, and I also gave up on the model.

But the carp surface is covered with the most beautiful scale patterning that strongly suggests a three-dimensional path curve pattern. What was now crucial was to discover whether or not this pattern continued smoothly over the dorsal and lower ventral edges without gaps and re-entrants. A closer look clearly showed a continuity of scales across the top of the body, and, though not quite as clearly, underneath. Some scales seem to double up interrupting the pattern somewhat.

The next step was to try to plot the supposed spirals in the front angled plane (through $P_1$) and the back angled plane (through $P_2$). Figure 14.29 is a summary sketch of this intent. What path curves in the four planes of the tetrahedron determine these spirals? There is a strong hint of this in the sketch where the estimated points (shown as red triangles) appear to progress around the head point, $P_1$, when they are linked together. This was a good first step and hinted at some kind of spiralling.

From the photograph I estimated there were about 30 scales around the body of the carp, and I assumed that they were spaced at equal angles, making them 12° apart. Could one establish a point in the front angled plane for each of these positions, and would it coincide with some spiral, even an asymmetric one? The kind of spiral I was looking for was like that in Figure 14.30.

I tried a number of possibilities. To plot the points the drawing paper had to be extended and almost covered our entire dining table. An averaged asymmetric spiral in the middle of an envelope of possibilities was estimated (heavy line in red in Figure 36.36). But this process did not work very well.

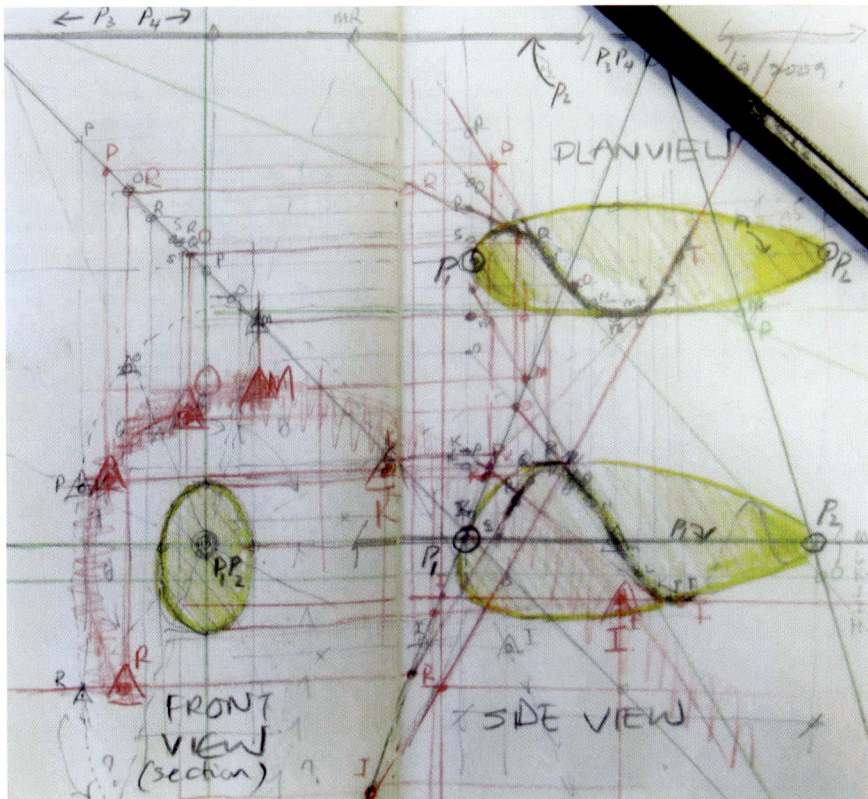

Figure 14.30

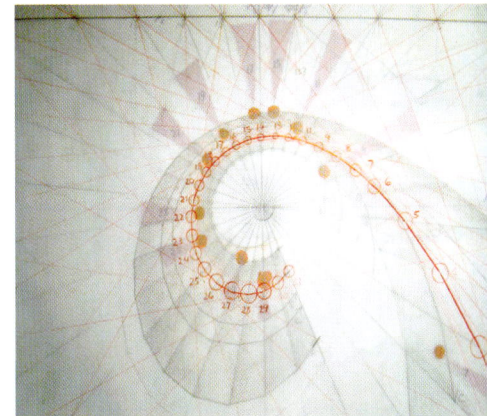

The constructions from the asymmetric spiral chosen did not coincide too well with the actual scale patterns. After very many trial spiroids I could still not get a good fit.

I thought it best to start over. I set up a vertical triangle but this time the line $p_2$ with a circling measure in it, not a step measure. This was an important change. Once again only every four fish scales were marked on the horizontal spinal line and a growth measure matched to these points. From this I could construct the idealised profile of the fish. It was not too bad a fit (Figure 14.32). The head and tail protrude, but that was expected.

On the front and back slanted lines in back and front planes there are two different growth measures. These different measures in the two slant lines help to determine the spiroids in the two front and back planes. I again estimated 30 scales per full cycle around the fish body, which would mean an angular spacing of 12°. A few test spiroids were drawn in, starting at the known projected point, but varying the angle, or slope across the plane (Figure 14.33).

As before the position of this top line was taken to be 300 mm

Figure 14.27 Pattern of carp scales over dorsal (top) edge
Figure 14.28 A similar pattern underneath
Figure 14.29 Sketch of the intended construct
Figure 14.30 An asymmetric spiral or spiroid
Figure 14.31 An averaged spiral within a (grey) envelope of possibilities

*Figure 14.32 Hypothetical profile in red*
*Figure 14.33 Some spiral lines drawn on the fish scales*

*Figure 14.34 The latest tetrahedron with two imaginary planes*
*Figure 14.35 A spiroid field generated by a non-equiangular circling measure*

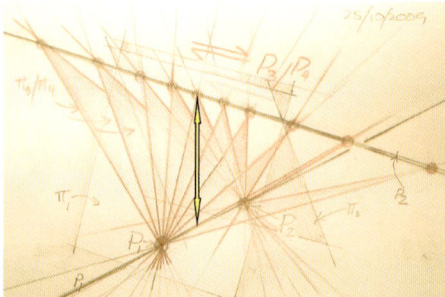

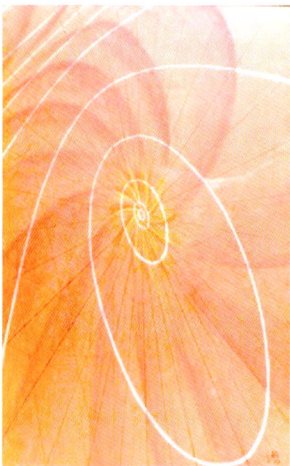

above the spinal line $p_2$ and perpendicular to it. Slowly the animal tetrahedron was being built with the help of an actual fish. It was now similar to our first animal tetrahedron in Figure 14.15 except that two planes were imaginary, rotating in opposing directions about the spinal line, $p_1$, determined by the circling measure of points, $P_3$ and $P_4$, in the top line $p_2$ (Figure 14.34).

The next step would be the acid test. Would I be able to find a plausible pair of counter rotating spiroids that gave, even approximately, the curves represented by the sequences of scales on the surface of the body of the carp? These two planar spiroids in the front and back planes will also determine the two point-wise spiroid cones in the two points, $P_1$ and $P_2$. It is the intersection of these two cones that should provide the track for the midpoints of the scales for one such sequence. If this is correct then all the adjacent sequences of scales should be able to be represented by as many nested spiroids as there are sequences (and in both directions if the fish skin is truly symmetrical bilaterally). From the data drawings, and when one full cycle appears to have been fulfilled, one should be able to count the number of intervening sequences that cover the central part of the body. I found there were 15 such sequences (Figure 14.33).

But I hit a snag. I tried dozens of spiroid drawings. None fitted. They were not too far out, but they simply did not accurately match the data points.

So what was wrong? Was the whole hypothesis flawed or mistaken? Had I missed something? Many, many sketches were done but the spiroid error envelope remained far too large and barely even included all the data points.

And then it struck me: one thing I had missed was that I assumed the spiroids were based on an equiangular spacing about the line $p_1$, the spine. But that was not necessarily so. I remembered I had once drawn a family of these, inspired by a drawing of Lawrence Edwards. At the time I drew it, it did not even occur to me that this might appear somewhere in nature.

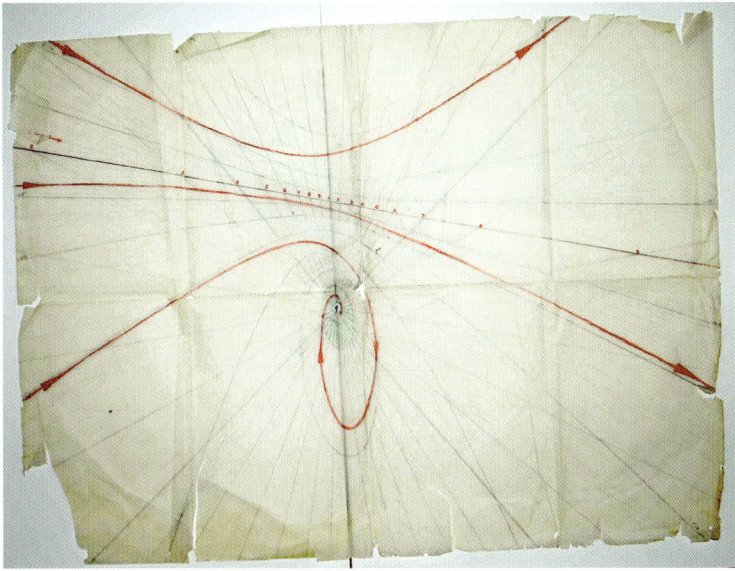

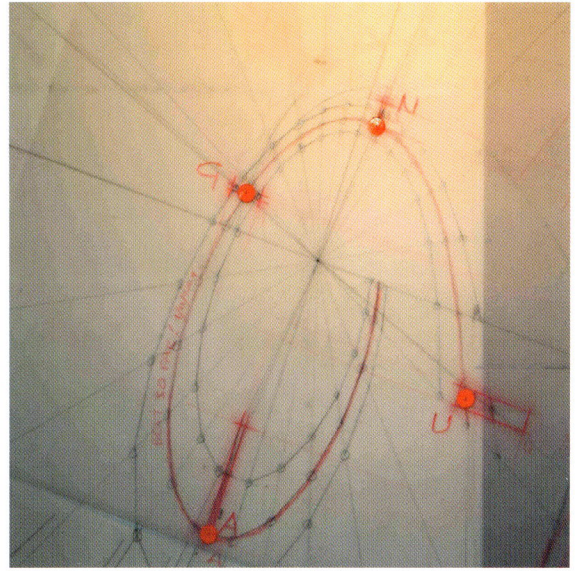

Figure 14.36 Original spiroid drawing
Figure 14.37 Spiroid in front plane

Circulating anticlockwise from the centre through some ellipses is a family of red spiroids (Figure 14.35). These forms do not depend on an equiangular circling measure of lines in the central point, but on a circling measure of lines in a point that is *not* necessarily equiangular. Once we have a circling measure of points in a line we can select any point outside the line and connect it to all the points and create a new circling measure of lines. I found my original drawing and have emphasised the spiroid curve, in red (Figure 14.36). Note that around the centre point the radiant lines are not equispaced, just as the points in the line are not equispaced either. I wondered whether the fish might like this idea.

Again I started with the data, using four of the significant data points. Using these it was possible to estimate the centre of this circling measure separately. Finally these new circling measure points were joined to the centre giving a new circling measure of lines. Now could one draw a spiroid through the data points?

The result I got after remarkably little experimentation was surprisingly good. This is shown as a red spiroid and going through the red dots representing the average data point positions (Figure 14.37). This was now good enough to see whether a spiroid could be found, using the same original data set. The data was used to project onto the back plane in exactly the same way as the projections onto the front plane. For the back or tail plane the drawings had to be yet larger, but it did not take long to get a reasonable fit for the projected points in the rear plane (Figure 14.38).

Finally, the early animal tetrahedron was developing well, with some of the path curve surfaces that control the main part of the fish

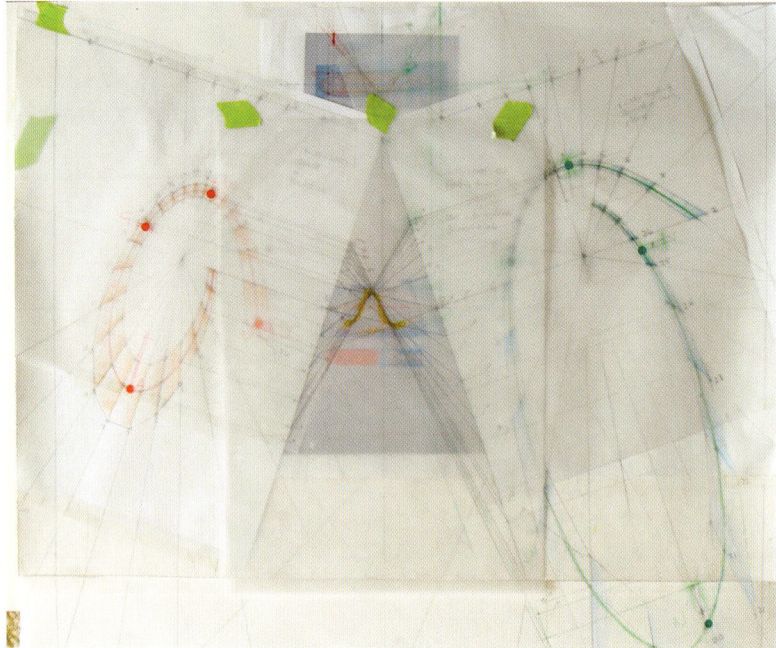

*Figure 14.38 Front and rear spiroids together*

body beginning to reveal themselves. What happens though when these two spiroids are made to interact with each other as they must do? For they are really part of two spiroid cones passing through points $P_1$ and $P_2$.

The final question is whether these two spiroids provide a curve of tangents (or points) which could reasonably be said to follow the scale sequence? A preliminary trial suggested the linking tangents (viewed from the side) looked decidedly good in a little more than just the middle of the fish body (Figure 14.39).

A refinement is to attempt to plot the points (not only lines) on the fish skin surface that represent an approximation to the path of the scale centres. This required a somewhat different approach but

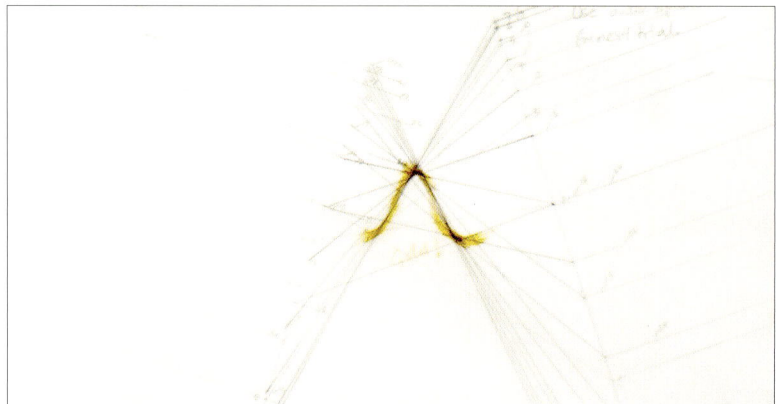

*Figure 14.39 Linking tangents (detail from Figure 14.38)*

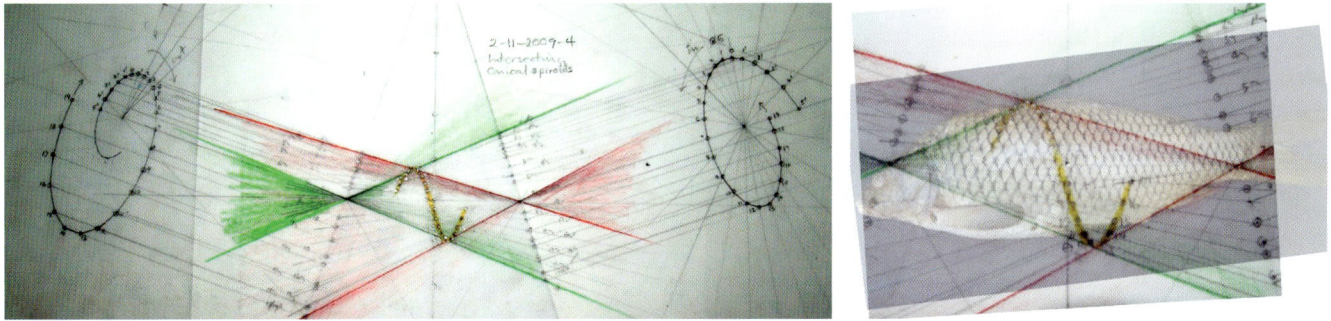

Figure 14.40 *Intersecting spiroid cones*
Figure 14.41 *Comparison of geometric curves and actual scale pattern*

using all the earlier constructed basis and data. In this case it was a question of finding the points coincident with the surface scales in the rotating planes shown in Figure 14.34. This required the drawing of another pair of spiroids, which were effectively the same proportions as those used to find the tangents, but considerably smaller. These two spiroids were linked, not this time by the tangents to the fish body but by two intersecting lines that give points on the surface of the fish body. The surface points are the result of intersecting lines on the surface of two intersecting spiroid cones (Figure 14.40). The overlay shows the fit is not perfect, but most of it is close enough to encourage further efforts (Figure 14.41).

## 14.5 Conclusion

So our hypothesis is that the main part of the fish form is embedded in a form field determined by this specific tetrahedral architecture. This third kind of tetrahedron, I believe, provides a basis for the overall initial form of the fish, an early creature in the animal world.

We saw that the tetrahedron governing the mineral word is a special case of the plant tetrahedron – all the lines are infinitely distant and the growth measure has become step measure. The plant tetrahedron in turn, is a special case of this more flexible animalic one. Only one line has become local, two others become imaginary and circling.

In the basic animal morphology it seems there is the intervention of something that allows an ever-increasing expression of sentiency or of soul life. In terms of evolution, early fish forms are generally long, thin and more circular in section, while later species are thinner and deeper. The evolutionary metamorphosis progressively gives expression to more subtle forms. The cross-section progressively alters from circular, to elliptical, to egg path curve, to re-entrant curves (like Cassini ovals). Is this a manifestation of orthogenetic intent – a guiding inner principle? Is the intent to eventually separate the head (by extending the neck stalk) and the four earth-bound limbs that ray from the reproductive area of the body?

# 15. Conclusion

## 15.1 Geometry in the human realm

When we first looked at the animal world, we saw one aspect that worked on the basis of plant geometry, and that was the egg. If we now look at the human realm, we find some of its forms or organs to some extent echo the egg form. Lawrence Edwards and Graham Calderwood have worked on the human heart form.

Observing the profile of the left ventricle of the heart, Edwards discovered that it followed a path curve. However, instead of the familiar bud or egg form which has a triangle with one point at infinity, so there are two parallel (horizontal) lines and one vertical line, the triangle for the path curves of the heart form have three local points (Figure 15.1). A full description can be found in Edwards, *The Vortex of Life*, Chapter 8. If we draw the three-dimensional equivalent of the heart profile then something like this stain glass panel emerges (Figure 15.2).

*Figure 15.1 The path curve corresponding to the heart profile with three local points (point Z is off the page to the left)*
*Figure 15.2 A stained glass panel indicating the heart form in three dimensions*

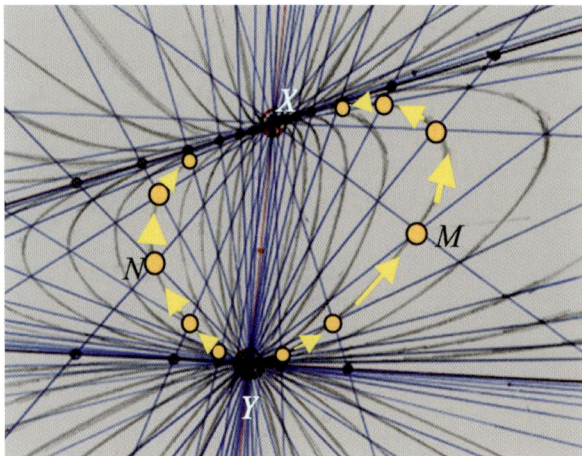

Despite initially seeking to find a form for the whole heart, Edwards had to resort to treating each ventricle separately. The muscle of the left ventricle of the heart is surprisingly thick around its centre: the right ventricle is much thinner, yet has a larger volume (Holdrege, *The Dynamic Heart and Circulation,* pp. 31f).

Incidentally, the brilliant nineteenth-century Scottish naturalist, James Bell Pettigrew illustrated the layers of the human heart. Most unusually, as an undergraduate, in 1860 he was invited to deliver the Croonian Lectures of the Royal Society and the Royal College of Physicians in London on the musculature of the heart. He discovered seven layers of muscle set at different angles. From the outer there were three layers spiralling anticlockwise, becoming progressively less steep, until the fourth layer was more or less horizontal; three inner layers spiralled clockwise becoming steeper (Figure 15.3).

The very early forms of the human cell also show signs of geometry. Soon after conception the egg cell starts to cleave or split. To begin with the overall size of the cell complex does not change. The cells simply get smaller. The first cleavage divides the cell into two giving bilateral symmetry. The second cleavage is perpendicular to the first and gives a total of four cells. The third cleavage cuts across this four-cell structure turning it into eight cells. There are now three mutually perpendicular planes of symmetry. There is almost something of the nature of mineral symmetry in that very early division that only lasts five or six days.

Lawrence Edwards also did some work on the path curves that form the later embryo. These again were an asymmetrical variation of the plant-like bud form modified by a vortex transformation.

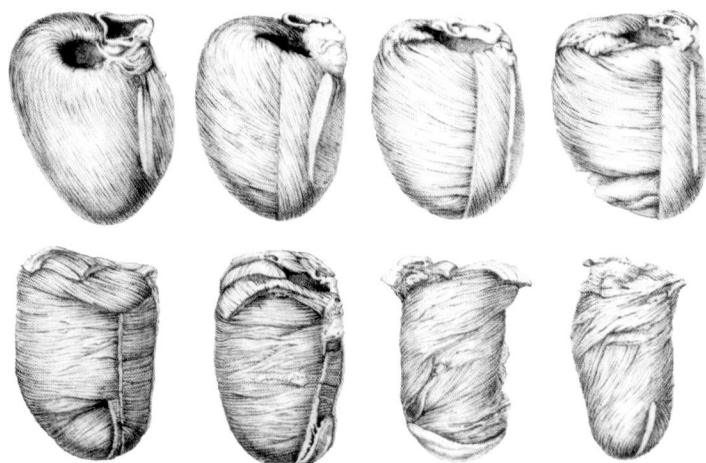

*Figure 15.3 Pettigrew's heart dissections (from* Design in Nature)

## 15.2 An overview of the geometry in different realms

In our considerations of the different kingdoms we have looked at the *orientation.* The mineral does not show a clear orientation, though there are aspects of horizontal layering visible on the large scale. The plant kingdom has a predominant verticality. The animal kingdom, whether fish or mammal or other kind, shows a dominant horizontal orientation. In the human realm we find a vertical orientation. This alone is something that sets the human apart from the animals and justifies looking at the human as a separate realm.

Each of the lines also displays a nodal pattern, a rhythm in the predominant line. The step measure giving equal sized crystals in the mineral, the more subtle growth measure we encountered in the spacing of stems and leaves on the plant or the vertebrae in the spine (not to mention the scale patterns of the fish), and no doubt there are similar patterns in the human spine.

Another aspect is the symmetries. The mineral displays a multitude of symmetries. In the plant these are reduced to rotational and bilateral symmetry. In the animal and human world the symmetry is further reduced to being just bilateral. No doubt there are small differences between left and right in the animal world, but it is characteristic of the human face that it is never *exactly* symmetrical. Internally, of course the organs of both animal and human are not totally symmetrical.

And in finding the tetrahedron that generates the path curves of the forms of these kingdoms we stepped from the infinitely large, but all-real, tetrahedron (for the mineral) to one that was partly infinite and partly local (for the plant world, with its vertical orientation). Then from this to a re-oriented tetrahedron that was wholly local (for the world of fish, with its horizontal orientation). The challenge of finding a tetrahedron that creates path curves, which match the human body as a whole, is truly great but I believe it would be again re-oriented vertically and all main line elements local. There may not even be such a tetrahedron. In any case an answer has so far eluded me.

## 15.3 Intelligent design?

We all take it for granted that there is an order to be found. Even scientists expect the mathematical/geometrical to be applicable. It is seen as an essential component of analysis. Geometry works because it is an inherent part of things. It only appears abstract as this is our current limited approach to it. We only sense the thinnest of abstractions, but that does not mean there is not a potent force at work in what the geometry represents.

A Catholic priest recently asserted that to call the majesty that meets us in nature merely 'intelligent', is to parody and belittle the innate wisdom in God's creation. 'Intelligent Design' is merely a sad diminutive, or a very low level acknowledgment of an extraordinary wisdom.

This extraordinarily intelligent design, this ineffable wisdom can be 'tested' using the insights of geometry. That has been my search here. My attempts have often been crude and I have only 'tested' a few species of the myriad spectrum of plants and creatures. If we develop a sense of awe and wonder for the world around us, we can begin to see that perhaps its evolution is not mere random chance, but has as its goal the upright human being. But for the moment we can leave it to biologists and evolutionary scientists to argue that point.

182

# Acknowledgments

Concerning those who have passed on: I owe a particular debt to Lawrence Edwards for showing me how it is possible to have a geometry of the plant world and teaching me personally many of the basics of this. Also to Roger McHugh, a man of penetrating insight from both left field and right. And to Rudolf Steiner for the only background story that made sense to me.

Overseas I owe much to many long phone sessions and visits to Nick Thomas in London, England, and visits and exchanges with Graham Calderwood in Aberdeen, Scotland. Also to the many people at the seminars in Strontian, Scotland, that we used to have, mentioning Lou de Boer, Ron Jarman and Stuart Brown in particular. And Simon Charter for the beautiful trout images he sent to me recently.

And locally, in Australia, David Bowden, Christel Post and Roger – all part of our one-time Morphics Group, Andrew Hill, Erik Thorvaldson, Peter Glasby, Fabiano Ximenes, Ron Vaisey (many, many discussions), Terry Forman, Garry Rollans, Marcel Maeder for conversations around the themes of my work.

Many past students of Glenaeon Rudolf Steiner School in Sydney who have kept my faith in future humanity and shown continued interest, among them Yasmin, Annika, Luke, Jenny, Paul, Monique, Daniel, Madelaine, Marco and Erik.

Many coffee shops of Sydney, especially at Pams in Castlecrag now with Lisa, Cornelius, Amelia and colleagues and Andronicus in Chatswood, Sydney, for putting up with me and supplying endless coffee.

Many museums, especially the Australian Museum in Sydney with its most excellent skeleton hall, and Mark McGrouther (Collection Manager, Ichthyology) for fishy discussions. Many gardens, especially the Royal Botanic Gardens in Sydney where

I have spent many hours (even years) seeking expressions of the formative architectures in the plant kingdom.

Rudie Kuiter, for his immediate help with images and descriptions of the fishy world. Ashley Miskelly, for his extraordinary collections and photos and beautiful publications on sea urchins. And to Jim Koulias, Jenny and Tim for giving me such a good price on some early published efforts of my work in notebook form.

And particularly to Christian Maclean at Floris Books for serious and thoughtful editing, and having the faith to make the book happen.

And to Norma, my wife, for her very concrete and continuous unstinting support, constantly challenging remarks and encouragements!

John Blackwood,
May 2012

# Bibliography

Abbott, Edwin A. *Flatland: A Romance in Many Dimensions*, Shambala, Boston and London 1999 (originally published 1884).

Adams, George, *Physical and Ethereal Spaces*, Rudolf Steiner Press, London 1965.

—, *Space and the Light of the Creation*, Published by the author, 1933.

Ayres, Frank, *Projective Geometry,* McGraw-Hill, New York 1967.

Baker, Douglas, 'A geometric method for determining shape of birds eggs', *The Auk,* 119 (4), 1179–86, American Ornithologists Union 2002.

Ball, Philip *The Self-Made Tapestry*, Oxford University Press, Oxford 1999.

Blatner, David, *The Joy of Pi,* Allen Lane, London 1997.

Bockemühl, Jochen, *Awakening to Landscape*, Natural Science Section, Goetheanum, Switzerland 1992

Bonewitz, Ronald Louis, *Rock and Gem*, Dorling Kindersley, London 2005.

Bortoft, Henri, *Goethe's Scientific Consciousness*, Institute for Cultural Research, 1986.

—, *The Wholeness of Nature: Goethe's Way of Science,* Floris Books, Edinburgh & Lindisfarne, New York 1996.

Casti, John L., *Five More Golden Rules*, John Wiley, New York 2000.

Church, A.H., *On the relation of phyllotaxis to mechanical laws,* Williams & Norgate, London 1904.

Clegg, Brian, *The First Scientist*, Constable, London 2003.

Colman, Samuel, *Nature's Harmonic Unity*, Benjamin Blom, New York 1971 (first published 1912).

Cook, Theodore Andreas, *The Curves of Life*, Dover, New York 1979 (first published 1914).

Critchlow, Keith, *The Hidden Geometry of Flowers,* Floris Books, Edinburgh 2011.

—, *Islamic Patterns*, Thames & Hudson, London 1976.

—, *Order in Space*, Thames & Hudson, London 1979.

—, *Time Stands Still*, Floris Books, Edinburgh 2007 (first published 1979).

Dennett, Daniel, *Darwin's Dangerous Idea*, Penguin, London 1995.

Doczi, Gyorgy, *The Power of Limits*, Shambala Publications, Colorado 1981.

Eberhart, Stephen, 'Grecian Amphorae as Path-Curve Shapes', *Mathematical Physics Correspondence,* Number 27, 1979.

Edwards, Lawrence, *The Field of Form*, Floris Books, Edinburgh 1982.

—, *Projective Geometry*, Floris Books, Edinburgh 2000.

—, *The Vortex of Life*, Floris Books, Edinburgh 2006 (first edition 1993).

—, *Supplements and Sequels,* www.vortexof life.org.uk/ reports.

Eisenberg, Jerome M., *Seashells of the World*, McGraw-Hill, New York 1981.

Gaarder, Jostein, *Sophie's World*, Phoenix House, London 1995.

Garland, Trudi Hammel, *Fascinating Fibonaccis*, Dale Seymour, New York 1987.

Ghyka, Matila, *The Geometry of Art and Life*, Dover, New York 1977 (first published in 1946).

Gleick, James, *Chaos*, Penguin Books, New York 1987.

Golubitsky, Martin and Stewart, Ian, *Fearful Symmetry*, Blackwell, Oxford 1992.

Goodwin, Brian, *How the Leopard Changed its Spots*, Weidenfeld and Nicolson, London 1994.

Gould, Stephen Jay, *I Have Landed*, Jonathan Cape, London 2002.

Hawking, Stephen, *The Universe in a Nutshell*, Bantam, London 2001.

Heath, Thomas L., *The Thirteen Books of Euclid*, Cambridge University Press, 1926.

Hitchens, Christopher, *God is not Great*, Allen & Unwin, New York 2007.

Hoffman, Paul, *The Man Who Loved Only Numbers*, Fourth Estate, London 1998.

Holdrege, Craig, *The Dynamic Heart and Circulation*, Association of Waldorf Schools of North America, Fair Oaks, USA 2002.

Huntley, H.E., *The Divine Proportion*, Dover, New York 1970.

Kandinsky, Wassily, *Point Line and Plane*, Dover, New York 1979 (first published 1926).

Kauffman, Stuart, *At Home in the Universe*, Oxford University Press, New York 1995.

Kepler, Johannes, *The Six Cornered Snowflake*, Paul Dry, Philadelphia 2010.

Klee, Paul, *The Thinking Eye*, Vol. 1, Lund Humphries, London 1961.

Koestler, Arthur, *The Sleepwalkers*, Penguin, London 1959.

Kollar, L. Peter, *Form*, privately published, Sydney 1983.

Kuiter, Rudie H., *Guide to Sea Fishes of Australia*, New Holland, Sydney 1996.

Livio, Mario, *The Golden Ratio*, Headline Review, London 2002.

—, *Is God a Mathematician?* Simon & Schuster, New York 2009.

Locher-Ernst, Louis, *Space and Counter-Space*, Association of Waldorf Schools of North America, Fair Oaks, USA 2003.

Lovelock, James, *The Ages of Gaia*, Oxford University Press, Oxford 1988.

Luminet, Jean-Pierre, *The Wraparound Universe*, Peters, Wellesley, USA 2008.

Mandelbrot, Benoit B., *The Fractal Geometry of Nature*, Freeman, New York 1977.

Maor, Eli, *The Story of a Number*, Princeton University Press, New Jersey 1994.

Marti, Ernst, *The Four Ethers*, Schaumberg Publications, Roselle, USA 1984.

Milne, John J., *An Elementary Treatise on Cross-Ratio Geometry*, Cambridge 1911.

Miskelly, Ashley, *Sea Urchins of Australia and the Indo-Pacific*, Capricornica, Sydney 2002.

Noble, Denis, *The Music of Life*, Oxford University Press, Oxford 2006.

Pakenham, Thomas, *Remarkable Trees of the World*, Weidenfeld & Nicolson, London 1996.

Peterson, Ivars, *Islands of Truth*, Freeman, New York 1990.

Pettigrew, J. Bell, *Design in Nature*, Longmans Green, London 1908.

Poppelbaum, Hermann, *Man and Animal*, Anthroposophical Publishing Company, London 1960.

—, *A New Zoology*, Philosophic-Anthroposophic Press, Dornach, Switzerland 1961.

Richter, Gottfried, *Art and Human Consciousness*, Anthroposophic Press, New York 1982.

Rohen, Johannes, *Functional Morphology: the Dynamic Wholeness of the Human Organism*, Adonis, New York 2007.

Romunde, Dick Van, *About Formative Forces in the Plant World*, Jannebeth Roell, New York 2001.

Ruskin, John, *The Elements of Drawing*, Dover, New York 1971 (originally published in 1857).

Saward, Jeff, *Labyrinth and Mazes*, Gaia Books, London 2003.

Schad, Wolfgang, *Man and Mammals*, Waldorf Press, New York 1997.

Schwenk, Theodor, *Sensitive Chaos*, Rudolf Steiner Press, London 1965.

Sheen, A. Renwick, *Geometry and the Imagination*, Association of Waldorf Schools of North America, Fair Oaks, USA, 1994.

Sheldrake, Rupert, *A New Science of Life*, Anthony Blond, London 1985.

Steiner, Rudolf, *Atomism and its refutation*, Article in 1890 Mercury Press, New York,.

—, *How can Mankind Find the Christ Again*, Anthroposophic Press, New York 1947.

—, *Karmic Relationships*, Vol. 1, Rudolf Steiner Press, London 2004.

—, *Man: Hieroglyph of the Universe*, Rudolf Steiner Press, London 1972.

—, *Mission of the Archangel Michael*, Anthroposophic Press, New York 1961.

—, *The Search for the New Isis, Divine Sophia*, Mercury Press, New York 1983.

Stevens, Peter S., *Patterns in Nature*, Penguin, New York 1974.

Stewart, Ian, *Does God Play Dice*, Allen Lane, London 1989.

—, *Life's Other Secret*, Allen Lane, London 1998.

—, *What Shape is a Snowflake?* Weidenfeld & Nicolson, London 2001.

Stockmeyer, E.A.K., *Rudolf Steiner's Curriculum for Waldorf Schools*, Steiner Waldorf Schools Fellowship, UK 1969.

Strauss, Michaela, *Understanding Children's Drawings*, Rudolf Steiner Press, London 1978.

Tacey, David, *The Spirituality Revolution*, Harper Collins, Sydney 2003.

Thomas, Nick, *Science Between Space and Counterspace*, Temple Lodge, UK 1999.

—, *Space and Counterspace: A New Science of Gravity, Time and Light*, Floris Books, Edinburgh 2008.

Thompson, D'Arcy Wentworth, *On Growth and Form*, Dover, New York 1992 (originally published 1916).

Tudge, Colin, *The Secret Life of Trees*, Penguin, London 2006.

Verhulst, Jos, *Developmental Dynamics in Humans and Other Primates*, Adonis Books, New York 2003.

Wachsmuth, Guenther, *The Etheric Formative Forces in Cosmos, Earth and Man*, New York 1927.

Whicher, Olive, *The Plant between Sun and Earth*, Rudolf Steiner Press, London 1952.

—, *Projective Geometry*, Rudolf Steiner Press, London 1971.

—, *Sunspace*, Rudolf Steiner Press, London 1989.

Wigner, Eugene, 'The Unreasonable Effectiveness of Mathematics in the Natural Sciences', in *Communications in Pure and Applied Mathematics*, Vol. 13, No. 1, February 1960.

Williams, Robyn, *Unintelligent Design*, Allen & Unwin, Sydney 2006.

Wolfram, Stephen, *A New Kind of Science*, Wolfram Media, Champaign, USA 2002.

Zajonc, Arthur, *Catching the Light, Bantam, New York 1993*.

# Index

# MORE BOOKS ABOUT GEOMETRY

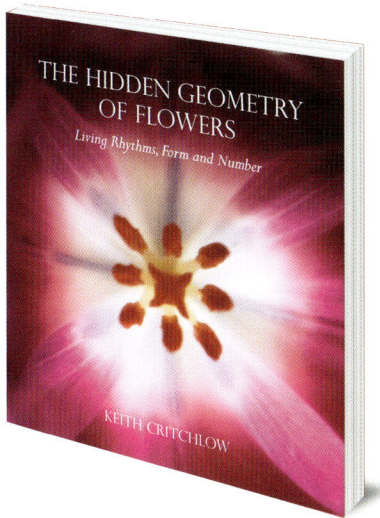

## The Hidden Geometry of Flowers: Living Rhythms, Form and Number

*Keith Critchlow*

*"This book is a definitive work on the geometry of the relationship between Nature and the Cosmos."* Z'ev ben Simn Halevi, Caduceus

A beautiful and original book in which renowned thinker and geometrist Keith Critchlow focuses on flowers as examples of symmetry and geometry. Fully illustrated with hand-drawn geometric patterns.

## The Vortex of Life : Nature's Patterns in Space and Time

*Lawrence Edwards*

*"An extraordinary book. His work will form one of the bases of our new holistic science."* Scientific and Medical Network Review

A pioneering work on bud and plant shapes, which indicates there are universal laws which guide an organism's growth into predetermined patterns.

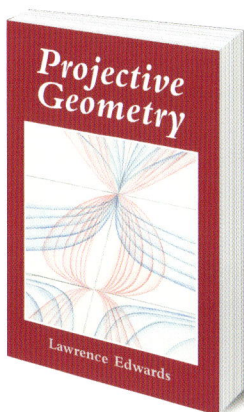

## Projective Geometry

*Lawrence Edwards*

*"Contains many geometrical illustrations throughout and can be approached by anyone willing to think and draw, without previous geometrical knowledge."* New View

A clear and artistic understanding of the qualities of projective geometry. Useful for high school Steiner-Waldorf teachers.

**www.florisbooks.co.uk**

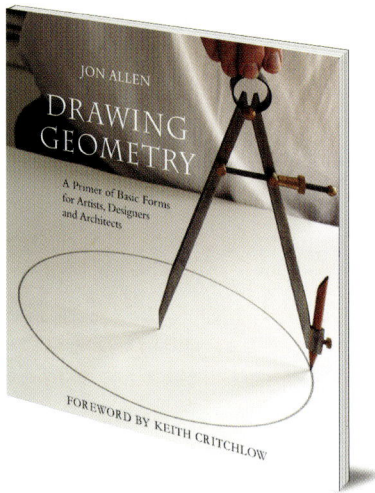

## Drawing Geometry:
## A Primer of Basic Forms for Artists, Designers and Architects

*Jon Allen*

*"A simple and beautiful book"* Scientific and Medical Network Review

Teaches professionals how to draw two-dimensional geometric shapes in simple step-by-step instructions.

## Making Geometry:
## Exploring Three-Dimensional Forms

*Jon Allen*

Teaches professionals how to make three-dimensional models of all the Platonic and Archimedian solids in step-by-step instructions.

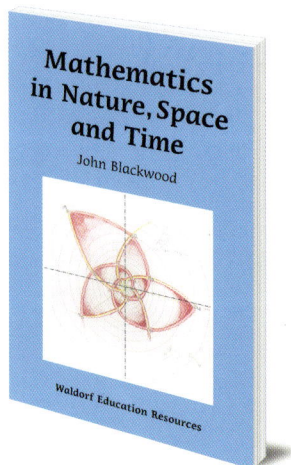

## Mathematics in Nature,
## Space and Time

*John Blackwood*

*"This is an exciting and unusual book"* Education Otherwise

A teachers' book for maths covering 'Mathematics in Nature', 'Pythagoras and Numbers', 'Platonic Solids' and 'Rhythm and Cycles'.

**www.florisbooks.co.uk**